IMPLEMENTING THE GOVERNMENT PERFORMANCE AND RESULTS ACT FOR RESEARCH

A Status Report

Committee on Science, Engineering, and Public Policy

National Academy of Sciences
National Academy of Engineering
Institute of Medicine

Policy and Global Affairs

NATIONAL ACADEMY PRESS
Washington, D.C.

NATIONAL ACADEMY PRESS • 2101 Constitution Avenue, NW • Washington, D.C. 20418

NOTICE: This volume was produced as part of a project approved by the Governing Board of the National Research Council, whose members are drawn from the councils of the National Academy of Sciences (NAS), the National Academy of Engineering (NAE), and the Institute of Medicine (IOM). It is a result of work done by a panel of the Committee on Science, Engineering, and Public Policy (COSEPUP). The members of the panel responsible for the report were chosen for their special competences and with regard for appropriate balance.

COSEPUP is a joint committee of NAS, NAE, and IOM. It includes members of the councils of all three bodies. For more information on COSEPUP, see www.nationalacademies.org/cosepup.

This material is based upon work supported by the National Science Foundation under Grant No. OIA-0073616, the National Research Council, the Department of Defense under Purchase Order SP4700-99-M-0510, the Department of Energy under Grant No. DE-FG02-00ER45803, the Department of Health and Human Services/National Institutes of Health under Contract No. N01-OD-4-2139, Task Order No. 60, and the National Aeronautics and Space Administration under Grant No. NASW-9937, Task Order No. 112. Additionally, this project, N01-OD-4-2139, Task Order No. 60, received support from the evaluation set-aside section 513, Public Health Service Act. Any opinions, findings, and conclusions or recommendations expressed in this material are those of the author(s) and do not necessarily reflect the views of the National Science Foundation, the Department of Defense, the Department of Energy, the Department of Health and Human Services, or the National Aeronautics and Space Administration.

International Standard Book Number: 0-309-07557-2

Implementing the Government Performance and Results Act for Research: A Status Report is available from the National Academy Press, 2101 Constitution Avenue, NW, PO Box 285, Washington, D.C. 20055 (1-800-624-6242 or 202-334-3313 in the Washington metropolitan area; Internet http://www.nap.edu).

Printed in the United States of America.

THE NATIONAL ACADEMIES

National Academy of Sciences
National Academy of Engineering
Institute of Medicine
National Research Council

The **National Academy of Sciences** is a private, nonprofit, self-perpetuating society of distinguished scholars engaged in scientific and engineering research, dedicated to the furtherance of science and technology and to their use for the general welfare. Upon the authority of the charter granted to it by the Congress in 1863, the Academy has a mandate that requires it to advise the federal government on scientific and technical matters. Dr. Bruce M. Alberts is president of the National Academy of Sciences.

The **National Academy of Engineering** was established in 1964, under the charter of the National Academy of Sciences, as a parallel organization of outstanding engineers. It is autonomous in its administration and in the selection of its members, sharing with the National Academy of Sciences the responsibility for advising the federal government. The National Academy of Engineering also sponsors engineering programs aimed at meeting national needs, encourages education and research, and recognizes the superior achievements of engineers. Dr. William A. Wulf is president of the National Academy of Engineering.

The **Institute of Medicine** was established in 1970 by the National Academy of Sciences to secure the services of eminent members of appropriate professions in the examination of policy matters pertaining to the health of the public. The Institute acts under the responsibility given to the National Academy of Sciences by its congressional charter to be an adviser to the federal government and, upon its own initiative, to identify issues of medical care, research, and education. Dr. Kenneth I. Shine is president of the Institute of Medicine.

The **National Research Council** was organized by the National Academy of Sciences in 1916 to associate the broad community of science and technology with the Academy's purposes of furthering knowledge and advising the federal government. Functioning in accordance with general policies determined by the Academy, the Council has become the principal operating agency of both the National Academy of Sciences and the National Academy of Engineering in providing services to the government, the public, and the scientific and engineering communities. The Council is administered jointly by both Academies and the Institute of Medicine. Dr. Bruce M. Alberts and Dr. William A. Wulf are chairman and vice chairman, respectively, of the National Research Council.

PANEL ON RESEARCH AND THE GOVERNMENT PERFORMANCE AND RESULTS ACT (GPRA) 2000

ENRIQUETA C. BOND *(Cochair)*, President, The Burroughs Wellcome Fund, Research Triangle Park, North Carolina
ALAN SCHRIESHEIM *(Cochair)*, Director Emeritus, Argonne National Laboratory, Argonne, Illinois
JOHN E. HALVER, Professor Emeritus in Nutrition, School of Fisheries, University of Washington-Seattle, Washington
BRIGID L. M. HOGAN, Investigator and Professor, Vanderbilt University Medical Center, Howard Hughes Medical Institute, Nashville, Tennessee
WESLEY T. HUNTRESS, JR., Director, Geophysical Laboratory, Carnegie Institution of Washington, Washington, D.C.
LOUIS J. LANZEROTTI, Distinguished Member, Technical Staff, Bell Laboratories, Lucent Technologies, Murray Hill, New Jersey
RUDOLPH A. MARCUS, Arthur Amos Noyes Professor of Chemistry, California Institute of Technology, Pasadena, California
STUART A. RICE, Frank P. Hixon Distinguished Service Professor, James Franck Institute, The University of Chicago, Illinois
HERBERT H. RICHARDSON, Associate Vice Chancellor of Engineering and Director, Texas Transportation Institute, The Texas A&M University System, College Station, Texas
MAX D. SUMMERS, Distinguished Professor, The Texas A&M University, College Station, Texas
MORRIS TANENBAUM, Retired Vice Chairman and Chief Financial Officer, AT&T, Short Hills, New Jersey
BAILUS WALKER, JR., Professor of Environmental and Occupational Medicine, Howard University, Washington, D.C.
ROBERT M. WHITE, University Professor and Director, Data Storage Systems Center, Carnegie Mellon University, Pittsburgh, Pennsylvania

Principal Study Staff:

DEBORAH D. STINE, Study Director and Associate Director, COSEPUP
SUSAN E. COZZENS, GPRA Consultant
DAVID M. HART, Case Study Consultant
ALAN ANDERSON, Consultant Science Writer
DAVID BRUGGEMAN, Intern, Christine Mirzayan Internship Program
CARL PICCONATTO, Intern, Christine Mirzayan Internship Program
REBECCA BURKA, Administrative Associate
KEVIN ROWAN, Project Assistant
NORMAN GROSSBLATT, Editor

COMMITTEE ON SCIENCE, ENGINEERING, AND PUBLIC POLICY

MAXINE F. SINGER *(Chair)*, President, Carnegie Institution of Washington, Washington, D.C.
BRUCE M. ALBERTS,* President, National Academy of Sciences, Washington, D.C.
ENRIQUETA C. BOND, President, The Burroughs Wellcome Fund, Research Triangle Park, North Carolina
LEWIS M. BRANSCOMB, Professor Emeritus, Center for Science and International Affairs, John F. Kennedy School of Government, Harvard University, Cambridge, Massachusetts
GERALD P. DINNEEN,* Vice President, Science and Technology, Honeywell, Inc. (retired), Edina, Minnesota
JAMES J. DUDERSTADT, President Emeritus and University Professor of Science and Engineering, Millennium Project, University of Michigan, Ann Arbor, Michigan
MARYE ANNE FOX, Chancellor, North Carolina State University, Raleigh, North Carolina
RALPH E. GOMORY, President, Alfred P. Sloan Foundation, New York, New York
RUBY P. HEARN, Senior Vice President, The Robert Wood Johnson Foundation, Princeton, New Jersey
SAMUEL H. PRESTON, Dean, University of Pennsylvania School of Arts and Sciences, Philadelphia, Pennsylvania
KENNETH I. SHINE,* President, Institute of Medicine, Washington, D.C.
EDWARD H. SHORTLIFFE, Professor and Chair, Department of Medical Informatics, Columbia University, New York, New York
HUGO F. SONNENSCHEIN, Charles H. Hutchenson Distinguished Service Professor, University of Chicago, Chicago, Illinois
PAUL E. TORGERSEN, John W. Hancock, Jr. Chair and President Emeritus, Virginia Polytechnic Institute and State University, Blacksburg, Virginia
IRVING L. WEISSMAN, Karele and Avice Beekhuis Professor of Cancer Biology and Professor of Pathology, Stanford University School of Medicine, Palo Alto, California
SHEILA E. WIDNALL, Abby Rockefeller Mauze Professor of Aeronautics, Massachusetts Institute of Technology, Cambridge, Massachusetts
WILLIAM JULIUS WILSON, Lewis P. and Linda L. Geyser University Professor, Harvard University, Cambridge, Massachusetts
WILLIAM A. WULF,* President, National Academy of Engineering, Washington, D.C.

Staff

RICHARD E. BISSELL, Executive Director
DEBORAH D. STINE, Associate Director
MARION RAMSEY, Administrative Associate

*Ex officio member.

PREFACE

In February 1999, the Committee on Science, Engineering, and Public Policy (COSEPUP) released a report titled *Evaluating Federal Research Programs: Research and the Government Performance and Results Act* (see Appendix E). The report recommended a set of criteria by which federal agencies might evaluate their programs of research in science and engineering. The criteria were intended to help agencies to respond to the Government Performance and Results Act (GPRA), enacted in 1993 (see Appendix F).

The National Academies were later asked by Congress to undertake another study, as part of the 1999 VA-HUD Independent Agencies Authorization Act, titled "Accountability of Federally Funded Research." Because many of the issues raised by Congress were addressed by COSEPUP in the original study, the Academies worked with the White House Office of Science and Technology Policy (OSTP) as indicated in the legislation to craft a study that would be most useful to all involved.

In a letter dated April 6, 1999, Dr. Neal Lane, director of OSTP, asked the Academies to undertake a more in-depth study of the actual application of GPRA to research programs as the agencies were shortly to release their first performance reports under GPRA. The study plan was endorsed by the House Committee on Science and by Senators William Frist, John Rockefeller, Jeff Bingaman, and Joseph Lieberman who were cosponsors of the original legislation. The specific charge to the panel was as follows:

As requested by Congress and the White House Office of Science and Technology Policy, this study would assist federal agencies in crafting plans and reports that are responsive to the Government Performance and Results Act (GPRA), OMB Guidance, and agency missions. The study would undertake independent assessments via case studies of the strategic and performance plans federal agencies have developed and of the responsiveness of their performance reports (which are due in March 2000) to the Government Performance and Results Act.

The assessment would take into account the agencies' missions and how science and technology programs and human resource needs are factored into agency GPRA plans. In addition, the study would suggest specific applications of recommendations from COSEPUP's earlier report entitled "Evaluating Federal Research Programs: Research and the Government Performance ad Results Act." In addition, workshops would be conducted where the agencies could share best practices regarding their performance reports and stakeholders views could be heard.

The Senators also requested that the Academies evaluate the extent to which independent merit-based evaluation achieves the goal of eliminating unsuccessful or unproductive programs and projects and to investigate and report on the validity of using quantitative performance goals for administrative management of these activities. COSEPUP decided not to pursue these analyses for the time being and to instead focus on the task above.

The National Academies formed the Panel on Research and the Government Performance and Results Act 2000 under the auspices of COSEPUP to respond to the request. This panel, which we chair, began its work by examining the GPRA performance reports each federal agency released in March of 2000. These performance reports provided the public with the first opportunity to see the implementation of GPRA.

In May, project staff at the behest of panel members met with the staff at 11 federal agencies to gain a better understanding of the methodology each used for their research programs. At this stage, problems with the charge to the panel emerged based on its discussions with the agency staff and the consultants and the panel's review of the agency performance plans.

Specifically, at its initial meeting in June, the panel members determined it was not appropriate to indicate the degree to which a given agency's work was acceptable, nor was it possible to conduct an in-depth review of each agency's program activities as would have been required to conduct an independent assessment of strategic and performance plans.

In the first instance, agencies were still in the experimental stage regarding the evaluation of research programs in response to GPRA. In the latter case, no single committee could mobilize the level of expertise necessary to conduct an in-depth review in the group of agencies selected given the tremendous diversity of the research programs each supported.

In sum, the panel determined it was not possible to provide the "independent assessment" of each agency's strategic and performance plan anticipated by Dr. Lane. In the spirit of the OSTP request, the panel instead decided to focus on the general methods and approaches each agency undertook. It also intentionally decided not to make agency-specific analyses beyond that which is presented in Appendix C summarizing each agency's approach.

Therefore, instead of attempting an investigation for which it was not equipped, the panel chose to take a "snapshot" of the current state of affairs of agencies' response to GPRA. After reviewing the process used by the 11 federal agencies, the panel in the end decided to select for review the five agencies that provide the most financial support for federal research programs. The five agencies selected were the National Science Foundation (NSF), National Institutes of Health (NIH), Department of Defense (DOD), Department of Energy (DOE), and National Aeronautics and Space Administration (NASA).

The panel then convened five focus groups—one on the process used by each agency—and a workshop to discuss overarching issues that affected all the agencies. Participants in the focus groups and the workshop included several panel members, members of agency scientific advisory groups, and staff from the agencies, Office of Management and Budget (OMB), General Accounting Office (GAO), and Congressional Research Service (CRS). Congressional committee staff were invited, but none attended. During each focus group, agencies were asked to respond to the following questions:

- What methodology is used for evaluating research programs under GPRA?
- What level of unit is the focus of the evaluation?
- Who does the evaluation of the research program under GPRA?
- What criteria are used for the evaluation?
- How are the selection and evaluation of projects related to the evaluation of the research program?
- How is the result communicated to different audiences (such as the S&T community, advisory committees, agency leadership, the administration, congress)?
- How is the result used in internal and external decision-making?

Their responses are summarized in Appendix C.

During the workshop, a number of overarching issues were discussed, including these:

- Criteria for evaluation.
- Aggregation of research programs for purposes of evaluation.
- Usefulness of GPRA.
- GPRA and the workload of agencies.
- Issues of timing.
- Verification and validation.

The results of the workshop are summarized in Appendix D.

The report itself should be considered a cross section or "snapshot" of agency responses to GPRA based on the agencies' own descriptions. We hope that the observations and recommendations presented here will be useful to other agencies in their efforts to implement GPRA and to oversight bodies in their efforts to supervise and facilitate the implementation. We believe, on the basis of first-hand observation, that the interactions during the focus groups and workshop were useful to all participants.

In the end, this panel does not attempt to recommend a single strategy to be used by all federal agencies in developing their plans to respond to GPRA. Instead, the panel, as requested by OSTP, has worked with individual agencies to focus on observations that could facilitate their responses to GPRA. Ideally, these lessons can be discussed and extended by all agencies and their oversight bodies to begin assembling agency-appropriate, broadly helpful strategies for GPRA compliance beyond that in COSEPUP's original report.

Enriqueta Bond
Alan Schriesheim
Panel Cochairs

ACKNOWLEDGMENTS

This report is the product of many individuals. First, we would like to thank all the agency, congressional, and White House staff and the agency advisory group members for their input (see Appendix C for focus group and workshop participants). The panel also extends thanks to Bob Simon and John Jennings, Senate staff, and Richard Russell and Beth Sokol, House staff, for their guidance. Without the help of all of them, the panel would have had great difficulty understanding the intricacies of the issues surrounding the implementation of GPRA for federal research programs.

Second, we would like to thank the reviewers of this report. This guide has been reviewed in draft form by individuals chosen for their diverse perspectives and technical expertise, in accordance with procedures approved by the National Research Council's Report Review Committee. The purpose of this independent review is to provide candid and critical comments that will assist the institution in making the published report as sound as possible and to ensure that the report meets institutional standards for objectivity, evidence, and responsiveness to the study charge. The review comments and draft manuscript remain confidential to protect the integrity of the deliberative process. We wish to thank the following individuals for their participation in the review of this report:

JOHN F. AHEARNE, Director, Ethics Programs, Sigma Xi, The Scientific Research Society, Research Triangle Park, North Carolina

EUGENE W. BIERLY, Senior Scientist, American Geophysical Union, Washington, D.C.
NICHOLAS P. BIGELOW, Professor, The University of Rochester, New York
RADFORD BYERLY, JR., Boulder, Colorado
THEODORE J. CASTELE, Fairview Park, Ohio
MELANIE C. DREHER, Dean, College of Nursing, The University of Iowa, Iowa City, Iowa
DAVID W. ELLIS, President and Director, Museum of Science, Boston, Massachusetts
FRANCIS B. FRANCOIS, Consultant, Bowie, Maryland
CUTBERTO GARZA, Vice Provost, Cornell University, Ithaca, New York
VICTORIA FRANCHETTI HAYNES, President, Research Triangle Institute, Research Triangle Park, North Carolina
SCOTT RAYDER, Director of Government Affairs, Consortium for Oceanographic Research and Education, Washington, D.C.
MICHAEL J. SAILOR, Professor of Chemistry, University of California, San Diego, La Jolla, California

Although the reviewers listed above have provided many constructive comments and suggestions, they were not asked to endorse the conclusions or recommendations, nor did they see the final draft of the report before its release. The review of this report was overseen by William G. Howard, an independent consultant in Scottsdale, Arizona. Appointed by the National Research Council, he was responsible for making certain that an independent examination of this report was carried out in accordance with institutional procedures and that all review comments were carefully considered. Responsibility for the final content of this report rests entirely with the authoring committee and the institution.

Finally, we would like to thank the staff for this project, including Deborah Stine, associate director of COSEPUP and study director; Alan Anderson, consultant writer, who worked with the panel to develop the text of the guide; Susan Cozzens and David Hart, who also served as our GPRA and case study consultants, respectively, on the report; Rebecca Burka, administrative associate, and Kevin Rowan, project assistant, who provided project support; David Bruggeman and Carl Picconatto, interns who provided research support; Norman Grossblatt, editor; and Richard Bissell, executive director of COSEPUP.

CONTENTS

EXECUTIVE SUMMARY

The Government Performance and Results Act (GPRA), enacted by Congress in 1993, requires that all federal agencies evaluate and report on the results of their activities annually.

Evaluating federal research programs in response to GPRA is challenging because we do not know how to measure knowledge while it is being generated, and its practical use might not occur until many years after the research occurs and cannot be predicted. For example, today's global positioning system is the result of research conducted 50 years ago in atomic physics. In 1999, the National Academies Committee on Science, Engineering, and Public Policy (COSEPUP) addressed this issue for research programs in its report *Evaluating Federal Research Programs: Research and the Government Performance and Results Act.* That report indicated that federal research programs could be evaluated by a process it called expert review that makes use of three evaluation criteria: quality, relevance, and leadership. Expert review is more than traditional peer review by scholars in the field. It also includes the users of the research, whether they are in industry, nongovernment organizations, or public health organizations or are other members of the public who can evaluate the relevance of the research to agency goals.

This followup report, by the COSEPUP Panel on Research and the Government Performance and Results Act 2000, describes the panel's analysis of how federal agencies that support science and engineering research are responding to GPRA. The panel decided to focus its work on the five agencies that provide the majority of federal funding for research: National Science Foundation (NSF), National Institutes of Health (NIH), Department of Defense (DOD), Department of Energy (DOE), and National Aeronautics and Space Administration (NASA).

As it began its examination of the strategic and performance plans and reports of these agencies, the panel found that, given the preliminary state of change of the agency's approach to GPRA for its research programs and the different organization and methodology of each, the panel could only conduct a "snapshot" of each agency's approach. Further, only general, not agency-specific, conclusions and recommendations were appropriate at this time. After a series of focus groups, a workshop, and numerous other communications with agency representatives and oversight bodies,[1] the panel reached the following 10 conclusions:

Conclusion 1: All five agencies have made a good-faith effort to develop reporting procedures that comply with the requirements of GPRA. Some agencies stated that GPRA compliance has added substantially to the cost of their planning and evaluation activities in the form of staff time and resources. Others report that they have been able to integrate GPRA with their traditional budget and planning processes although at some cost of time and effort.

Conclusion 2: Some agencies are using the GPRA process to improve their operations. These agencies report benefits in strengthening program management and enhancing communication about their programs to the users of research and the general public. The need to do so depends on the goal of that agency and the degree to which there is concern about a given field of research or about new and emerging programs. A few agencies

[1]Primarily Congress's General Accounting Office and the White House Office of Management and Budget.

found that GPRA requirements added to their reporting workload and are still struggling to adapt to these requirements.

Conclusion 3: The most effective technique for evaluating research programs is review by panels of experts using the criteria of quality, relevance, and, when appropriate, leadership. Agency approaches to GPRA research programs demonstrate the utility of expert review using the same criteria of quality and relevance as outlined in COSEPUP's original report. The international leadership criterion is generally not evaluated by most federal agencies at this time, although several are interested in such a measure. However, given the diversity in mission, complexity, culture, and structure of federal agencies that support research, it is not surprising that their approaches to GPRA have varied. One size definitely does not fit all.

Conclusion 4: Oversight bodies and some agencies need clearer procedures to validate and verify agency evaluations. In particular, oversight bodies expressed a desire for better understanding of the methodology and results of expert review evaluations.

Conclusion 5: Agencies choose to aggregate their research programs at different levels. Some agencies provide evaluations on a field-specific or program-specific basis; others do so for the research program in its entirety. Aggregating at a high level can make it difficult for oversight bodies to clearly see and understand the method and programs that are the focus of the analyses.

Conclusion 6: The development of human resources as an agency objective sometimes does not receive explicit emphasis or visibility in GPRA plans and reports.

When this objective is explicit, it affirms the value of educating young scientists and engineers by involving them in the research programs of their advisers. In addition, such an explicit linkage between research and education makes it easy to show how reductions in research funding can jeopardize the preparation of the scientists and engineers the nation will need in the future.

Conclusion 7: Agencies often receive conflicting messages from oversight bodies about the desired format, content, and procedures to be used in GPRA compliance. For example, one agency made an effort to tie its GPRA reports more closely to its annual budget, as required in the act, only to be told by a congressional committee to return to a previously used format; another was told the reverse.

Conclusion 8: Due to timing requirements built into the legal guidelines of GPRA, agencies find that they must begin work on performance plans before the relevant performance reports are complete. As a result, the potential benefit of GPRA in providing a mechanism for incorporating performance results of previous years into performance plans for later years is limited. A longer performance schedule—say, 3 years—would probably provide sufficient timing for most cases.

Conclusion 9: Communication between agencies and oversight groups is not sufficiently regular, extensive, or collaborative. During focus groups, the workshop, and interviews, it was consistently clear that improved communication between these two sectors could reduce the difficulties and misunderstandings experienced by some agencies.

Conclusion 10: The degree to which the results of GPRA reporting of research programs are being used

by oversight groups for programmatic decision-making is not clear. In particular, agencies have not yet seen the use of their reports in the congressional decision-making that determines the size and priorities of their budgets.

On the basis of these observations, the panel offers specific recommendations in Chapters 2 and 3. They can be summarized in the form of the following four general recommendations:

Recommendation 1: Federally supported programs of basic and applied research should be evaluated regularly through expert review, using the performance indicators of quality, relevance, and, where appropriate, leadership.

Recommendation 2: Agencies should continue to improve their methods of GPRA compliance and to work toward the goals of greater transparency, more-realistic reporting schedules, clear validation and verification of methods, and the explicit use of human-resources development as an indicator in performance plans and reports.

Recommendation 3: Agencies and oversight bodies should work together as needed to facilitate agencies integrating their GPRA requirements with their internal planning, budgeting, and reporting processes. In addition, they should work together to adjust the timing of GPRA reporting to capitalize on the value of the planning process.

Recommendation 4: Agencies should strive for effective communication with oversight groups on the implementation of GPRA. For their part, oversight

bodies should clarify their expectations and meet more often among themselves to coordinate their messages to agencies.

Much has been learned about the procedures of planning, evaluation, and management in the last several years, and some value will have been gained by the agencies from their own discussion of accountability. However, one key remaining question is the degree to which oversight groups are using the results of the "results act" for programmatic decision-making. Unless the agency responses to GPRA are useful to Congress in the urgent task of setting priorities and budgeting, the value of the act might not warrant the time and effort it requires of the federal government. But by working more closely than they have in the past, the federal agencies and the oversight bodies can implement the letter and spirit of GPRA in ways that lead to greater efficiency, lower cost, and more-effective research programs that are demonstrably conducted in the national interest.

CHAPTER 1

THE CHALLENGE OF EVALUATING RESEARCH

Passage of the Government Performance and Results Act (GPRA) in 1993 reflected a desire on the part of the public and their representatives in Washington for more effective and efficient use of public funds. GPRA requires a heightened degree of accountability in the planning, performance, and review of all federally funded activities.

The fraction of the United States budget invested in scientific and engineering research is relatively small, but it is highly visible, extremely important to the nation's future, and subject to lively debate. Federal funds support a total of some $20.2 billion[1] worth of basic research in 1998; about half that amount goes to the National Institutes of Health (NIH).[2] About $50 billion more is spent on applied research and development, of which a large portion is devoted to the procurement and testing of weapons systems. In all, the public investment in defense, health care, environment, space exploration, and other research-based endeavors constitutes a substantial public commitment.

In return for that investment, the public rightly expects substantial returns in the form of recognizable and useful outcomes. GPRA, as applied to scientific and engineering research, translates that expectation into a requirement for regular evaluations of

[1]National Science Board. 2000. *Science and Engineering Indicators*. Text Table 2-1.

[2]The next-largest recipients are the National Aeronautics and Space Administration (NASA), the Department of Energy (DOE), and the National Science Foundation (NSF), which are each allocated about 12% of federal funding for basic research.

Basic Research and Applied Research

As a search for the unknown whose outcomes are virtually unlimited, research defies exact definition. Intellectually, it is apparent that the performance of research takes place across a continuum of thought and action, from the abstract reasoning of a single individual to a multi-billion-dollar program of technological complexity, such as a mission to Mars.

However, to satisfy administrative or intellectual needs, it has often been convenient to separate "basic" research from "applied" research. In that spirit, basic research is often thought of as an unfettered exploration of nature whose only required output is new knowledge and whose outcomes are unknowable in advance. Applied research might be described as an activity whose outputs are also new knowledge, but knowledge whose nature and use are explicitly needed to achieve a specific useful outcome.[3]

Any research process is complex and has many feedback loops. A question raised during "applied" research might kindle a "basic" question that leads to new fundamental understanding. The knowledge "output" of basic research might—often after years or even decades—find utility as a practical "outcome." For example, some of Louis Pasteur's most fundamental understandings about microbiology grew out of practical attempts to control spoilage in beer and wine. In contrast, a knowledge-seeking study in basic research can lead to a discovery of great practical value. The atomic phenomenon of stimulated emission identified by Einstein in 1917 led eventually to the laser light that carries our e-mail today along fiber-optic lines.[4]

In managing and funding research, it is important to understand the open-ended possibilities of any research activity, no matter how it is categorized, and to encourage the freedom of inquiry that leads beyond what is already known. Any imagined distinctions between "basic" and "applied" research are less important than this unimpeded freedom to follow one's intuition and evidence in the service of improved understanding. In practice, research managers must have the insight to balance the need for predictable results with the desire for unexpected breakthroughs.

[3]For example, a research effort to make an amplifier by using semiconductors did not succeed. It was suggested that something might be happening on the surface of the semiconductor that interfered with the desired result. A basic study of the semiconductor surface began, which led to the discovery of the transistor effect.

[4]The National Academies publish *Beyond Discovery: The Path from Research to Human Benefit*, a series of articles that describe applications of basic research that could not have been anticipated when the original research was conducted. The series, published four to six times per year, is available on the National Academies Web site, www.nationalacademies.org/beyonddiscovery.

federal research program performance and public disclosure of the results.

Similarly, the public's representatives in Congress expect from agencies a sufficiently clear explanation of agencies' research activities to allow them to set priorities and manage agency budgets. Congress's desire for simplified and understandable information about research programs is reflected in the act's requirement of planning and reporting mechanisms.

Federal agencies that support research have moved by stages toward full implementation of GPRA over the last 4 years, with the central objective of providing a regular accounting of their research activities. They have spent substantial staff time designing ways to adapt their procedures to the act and have provided extensive plans and reports about their procedures and achievements (see Appendix G).

Nonetheless, both the agencies and oversight bodies have wrestled with interpreting, implementing, and communicating about GPRA. This report attempts to examine the agencies' progress toward meeting objectives, discusses some of the problems encountered, and recommends several actions intended to benefit all parties.

Because of the complexity of responding to GPRA and because the methods used by federal agencies are still in early stages of development, the panel decided to focus its effort on creating an accurate picture of the processes being developed rather than on its specific mechanisms.

To achieve that, the panel used a series of focus groups in which the agencies shared their experiences in creating their performance plans and reports, and representatives of oversight bodies provided their perspective and interacted with agency representatives. The focus groups were followed by a workshop and supplemented by numerous interviews with agency personnel and oversight groups. Of 11 agencies that support research in science and engineering, five were chosen for in-depth examination: the

Department of Defense (DOD), Department of Energy (DOE), National Aeronautics and Space Administration (NASA), National Institutes of Health (NIH), and National Science Foundation (NSF). Together, these five agencies account for some 94% of the federal government's spending on basic research.

The remainder of this chapter summarizes COSEPUP's first report on the issue of evaluating federal research programs. This report entitled, *Evaluating Federal Research Programs: Research and the Government Performance and Results Act* recommends that federal research programs be evaluated using a process called expert review and the criteria of quality, relevance, and leadership.

1.1 Barriers to Evaluating Research and the Solution

The difficulty of using measurements to evaluate research arises because the purpose of research is to provide knowledge and better understanding of the subject under study. For example, research in physics is aimed at a better understanding of the laws of nature that govern the behavior of matter and energy. A specific case is research into those materials that become superconducting at low temperatures. The eventual outcome of such work might be knowledge about synthesis of materials that are superconducting at room temperature. Practical outcomes would be new classes of electronic devices and high-efficiency motors and power-transmission systems. However, those outcomes might not occur for many years. Indeed, research might demonstrate that such materials cannot be made—also a valuable result that would save us from the futile pursuit of such outcomes.

Because we do not know how to measure knowledge while it is being generated and when its practical use cannot be predicted, the best we can do is ask experts in the field—a process called *expert review*—to evaluate research regularly while it is in progress. These experts, supplemented by quantitative methods, can determine whether the knowledge being generated is of high quality, whether

Terms of the Government Performance and Results Act

GPRA requires agencies to produce three documents: a strategic plan, a performance plan, and a performance report. A strategic plan must cover a period of at least 5 years and be updated every 3 years. The performance plan and performance report must be submitted annually.

The performance plan must list specific performance goals for the fiscal year of the budget it accompanies. Agencies are required to relate their performance goals to the broader objectives of the strategic plans and to specific activities described in the annual agency budget. The plans must establish performance goals for each program activity, and these goals must be expressed in an "objective, quantifiable, and measurable form." The performance report is intended to be included in each agency's "accountability report," due 6 months after the end of the fiscal year.

For many government activities—such as the provision of benefits to a segment of the population, the construction of a highway, or the implementation of a new service—the setting of performance goals and the annual assessment of progress are conceptually straightforward. That is, they are able to list their performance goals in quantifiable terms and report on their progress toward those goals by using specific metrics and time lines.

For research activities in science and engineering, however, especially those involving basic research, it is difficult or impossible to know the practical outcomes of activities in advance or to measure their progress annually with quantifiable metrics or milestones. Although it is desirable to use traditional measures of scientific excellence—including publications in refereed journals, frequency of citations, patents, honors and awards from professional associations—such measures apply most usefully to individuals rather than groups, and they offer only limited perspective on the likely outcome of entire programs. The difficulty of predicting outcomes presents challenges both to agencies whose primary mission is research, such as NSF and NIH, and to the research components that are often relatively small parts of mission agencies. The deeper reason for the difficulty is embedded in the nature of research itself, as discussed in the box on "Basic Research and Applied Research."

Accordingly, the act allows an "alternative form," as approved by the OMB, for agencies that do not find it feasible to express their performance goals in quantitative form. A number of agencies have experimented with alternative forms, with mixed reviews on achieving GPRA requirements. Agencies are still seeking effective response mechanisms that both they and oversight groups find useful. To a large extent, the primary source of difficulty is the complex nature of research itself.

it is directed to subjects of potential importance to the mission of the sponsoring agency, and whether it is at the forefront of existing knowledge—and therefore likely to advance the understanding of the field.

Expert review is a well-understood and widely applied technique that is used by congressional committees, in various other professions, by industry boards, and throughout the realm of science and engineering to answer complex questions through consultation with expert advisers. Virtually all science and engineering programs in federal agencies, universities, and private laboratories use at least some expert review to assess the quality of programs, projects, and researchers. Expert review is more than traditional peer review by scholars in the field. It also includes the users of the research, whether they are in industry, nongovernment organizations, or public health organizations or any other members of the public who can evaluate the relevance of the research to agency goals.

This report does examine other mechanisms for analyzing research, including bibliometric analysis, economic rate of return, case studies, and retrospective analysis. All methods were found to have some utility, but the people best qualified to evaluate any form of research are those with the knowledge and experience to understand its quality, relevance, and leadership and, in the case of applied research, its application to public and agency goals.

Furthermore, in many research programs, progress toward outcomes is not reflected in outputs that can be measured in a single year. In such cases, the value of the work might appear as an accumulation of discrete steps or sometimes abrupt insights that require two, three, or even more years to emerge. So a retrospective analysis over a number of years is necessary. For other research programs, progress toward specified practical outcomes can be measured annually with milestones and other quantitative approaches common in industry and some parts of the federal government.

For any long-term research program, results can be described annually—given a clear understanding of the research process. In the example of the search for room-temperature superconductors, one might expect such first-year results as drafting a request for proposals, evaluating responses, and funding the best of them. The research results themselves would begin to emerge in the middle years of such a program, and the interpretation of results and outcomes would emerge in the last years and perhaps be accompanied by planning for more-distant outcomes. The point is that the process is distorted if one expects to evaluate only the research results of a program for any given year of a long-term effort.

1.2 COSEPUP's Evaluation Criteria

COSEPUP proposed three evaluation criteria that should be used during the expert review process: quality, relevance, and leadership. These are described in more depth below.

1.2.1 Quality. Review of the quality of research via peer review is the most common form of expert review. Peer review is applied throughout the scientific and engineering communities to the work of laboratories and individuals. All the agencies involved in the focus groups said that they use it to evaluate programs. Because one's professional peers are uniquely familiar with the standards, context, history, and trends of a field, they are uniquely qualified to assess the quality of a research endeavor and to recommend improvements.

The sine qua non of quality review is objectivity. Oversight agencies want more evidence that the personal connections or histories of reviewers do not influence their opinions of the institutions or individuals under review. That concern is legitimate and forms the basis of the custom of seeking out panels that are not only expert, but also independent, in a professional sense, of the object of review. Expert review must be carried out by individuals who have technical expertise in the subject being reviewed but who are professionally independent of the program under review. Although it is true that those who are qualified to do quality reviews have some loyalty to the field, their potential bias is balanced by the strong tradition of honesty in the review process.

1.2.2 Relevance. Relevance review is conducted by panels of expert peers joined by experts in related fields, potential users of the results of research, or other interested members of the public. Advisory committees are typically asked to answer the question, Does the agency's research address subjects in which new understanding could be important in fulfilling the agency's mission? The goal is to evaluate the relevance of a research program or project to the agency's goals. User communities are taken to consist of those for whom agency research is intended to be relevant, including members of the academic and private sectors. For example, federally supported health research is assumed to benefit patients, medical practitioners, pharmaceutical companies, and other groups that use the results of research to develop new therapies and new products and to reap the benefits of new cures. It is important that these users help to evaluate the research "product" they hope to use. At the same time, it is essential to choose user groups with care so that they understand the need for the community's broad interests and do not focus too narrowly on single issues.

1.2.3 Leadership. Review of leadership was proposed in the first COSEPUP report as a potentially effective evaluation criterion to test whether research is being performed at the forefront of scientific and technologic knowledge on an international level.

In its *Goals* report of 1993, COSEPUP wrote that for the sake of the nation's well-being, the United States should be among the leaders in all major fields of science and pre-eminent in selected fields of national importance. The rationale is that the nation must be performing research at the forefront of a field if it is to understand, appropriate, and capitalize on current advances in the field, no matter where they occur.[5] Review of leadership is a new but promising means to gauge the place of a nation's research programs.

Review can be accomplished by the technique of *international benchmarking;* an exercise carried out by a panel of non-US and US experts whose technical expertise and international perspective qualify them to assess the standing of a research program or an entire field. They are asked to assess the relative position of US research today, the expected relative position of US research in the future, and the key factors influencing relative US performance. The premise for using the leaders of a research field is that they are in the best position to appraise the quality of researchers in their field, to identify the most promising advances, and to project the status of the field into the future.

As an experiment, COSEPUP panels performed international benchmarking in three fields—mathematics, immunology, and materials science and engineering—and found it to be faster and less expensive than procedures that rely entirely on the assembly of quantitative information, such as numbers of dollars spent, papers cited, plenary lectures delivered at international congresses, and scientists supported.[6]

The panels also found good correlation between the qualitative judgments of experts and the results of quantitative indicators. In addition, panels concluded that quantitative measures by themselves are inadequate indicators of leadership, both because quantitative information is often difficult to obtain or compare across national borders. Also, quantitative information generally

[5]COSEPUP. 1993. *Science, Technology, and the Federal Government: National Goals for a New Era.*

[6]COSEPUP. 2000. *Experiments in International Benchmarking of US Research Fields.*

illuminates only a portion of the research process. In other words, numbers of papers, patents, or citations should be used as indicators of the generation of innovative technologies, but they do not by themselves necessarily illuminate the most promising or important activities in a field. An experiment in mathematics by NSF that produced results similar to COSEPUP's mathematics study also lends credence to the benchmarking technique despite differences in the makeup and mandates of the two panels.[7]

1.3 Organization of this Report

In Chapter 2 the panel provides its assessment of methods being used by agencies to comply with GPRA. Chapter 3 discusses some difficulties in communication between agencies, oversight bodies, and the public about GPRA. Chapter 4 provides the panel's general conclusions and recommendations.

[7]National Science Foundation, *Report of the Senior Assessment Panel of the International Assessment of the US Mathematical Sciences*, Arlington, VA, March 1998.

CHAPTER 2

AGENCY METHODS

This chapter examines the methods developed so far by agencies to evaluate their research programs, some of the difficulties encountered, and some features of the interactions between agencies and their oversight bodies[1] in both the legislative and executive branches. The observations here are based on the conversations of the panel with agency and oversight staff as well as with members of agency expert review panels in its focus groups and workshop. At the focus groups, the five agencies examined—DOD, DOE, NSF, NIH, and NASA—were asked to respond to the following questions regarding their methodology:

- What methodology is used for evaluating research programs under GPRA?
- What level of unit is the focus of the evaluation?
- Who does the evaluation of the research program under GPRA (e.g., advisory committee, staff, combination)?
- What criteria are used for the evaluation?
- How does the selection and evaluation of projects relate to the evaluation of the research program?

[1]The oversight bodies with primary responsibility for assisting agencies and evaluating their efforts to comply with GPRA are Congress and its General Accounting Office (GAO), with input from the Congressional Research Service (CRS), and the White House Office of Management and Budget (OMB), with input from the Office of Science and Technology Policy (OSTP).

It was apparent to the committee that all the agencies interviewed have made good-faith efforts to comply with the requirements of GPRA. During the focus groups, they described in detail the evolution of their approaches, their frequent midcourse corrections, and their expenses in time and effort. The act forbade the use of outside consultants or additional hiring to design and execute GPRA responses, and for most agency officials the demands of GPRA produced an increased workload that promises to continue for some time. (For more details, see the agencies' responses, summarized in Appendix C.)

2.1 Expert Review

All agencies use expert review panels to evaluate their research programs. However, in response to GPRA, each of the agencies addresses the issue of expert review in a different way. Further, while some have well-established procedures that they are just refining, others are still at the very early stages of development.

Both NSF and NIH use advisory groups who produce evaluations via an alternative format approved by OMB. Using this method, there is no attempt to quantify a goal or the degree to which it has been met. Rather, goals *are successfully met* or *substantially exceeded* in NIH's case or *successful* or *minimally effective* in NSF's case, as determined by a single (in NIH's case) or multiple (in NSF's case) expert review panels. At NIH, a single overarching panel evaluates all NIH's research programs at one time. At NSF, numerous committees of visitors review individual research programs on a rolling 3-year basis. The results of those evaluations are then provided to several advisory committees whose membership represent several disciplines.

DOD uses a process called Technology Area Reviews and Assessments (TARA) to evaluate science and technology programs through expert peer reviews. In the DOD process, basic research is not isolated from applied research and advanced technology development. All three categories—6.1 (basic research), 6.2 (applied research), and 6.3 (advanced development)—are evaluated as overlapping parts of the technology area under review, with clear links to what discoveries are expected.

In the case of both NASA and DOE, both generally conduct extensive peer review of their projects and programs using external advisory committees. However, each has faced difficulties in translating their existing activities into the GRPA process in terms of setting an appropriate level of unit for evaluation and for finding relevant performance measures. Some programs (called *enterprises* at NASA), such as that in Basic Energy Sciences at DOE and at NASA, have had more success than others within the agency. Both these agencies are undergoing major redesign efforts in how they respond to GPRA for their research programs.

In the case of all agencies, staff and advisory committee members expressed concern that GPRA-related activities diverted advisory committee members from their original activities or added new activities.

Furthermore, agency representatives expressed concerns that the balance of the existing membership might need to be modified. Although expert review panels often include members who are expert in fields "adjacent" to the field under review, GPRA documents reviewed by the panel did not clearly identify reviewers who were representatives of "potential user communities" or, where appropriate, the public. Nor do documents show how agencies attempt to determine the extent of their particular "potential user universe." Because agency research is supported wholly by public funds, it is appropriate during the review process to consider how the interests of users are served. In addition, explicit statements about the reasons for pursuing particular fields or programs could help agencies to focus on their most productive initiatives and avoid wasting resources on those with low potential. Further, international members from outside the United States would help respond to questions of international leadership.

Recommendation M-1

Agencies should continue to take advantage of their existing expert review panels, but should review the balance in their membership, particularly the need to include user groups, and the time panel members devote to GPRA versus other topics so that it is not excessive. In addition, they should review the degrees to which internal and external reviewers are used.

2.2 Evaluation Criteria

This section summarizes the degree to which the agencies are using COSEPUP's proposed criteria (quality, relevance, leadership) to evaluate their research programs.

2.2.1 Quality. According to the agencies themselves, quality is the most widely and traditionally used of the three criteria. By custom, the quality of research is evaluated by peer-review committees that include members both inside and outside the program under review.

In rare cases, agencies use internal reviewers or program monitors in place of external reviewers. That practice might be deemed necessary when those best qualified to perform evaluations work in the same agency, although perhaps in a different division. In such cases, the independence, rather than the external position, of the reviewer(s) is judged to be a validating factor, and the degree of independence is confirmed by agency administrators. For example, a program monitor performs evaluations of the electrochemistry program in the Office of Naval Research by meeting annually with grantees; the officer is uniquely familiar with the details of the research. In other cases, security or other considerations might dictate a need for internal review. In general, however, review by outside experts is preferred.

Recommendation M-2

Agencies should continue to use peer review to evaluate the quality of their research programs.

2.2.2 Relevance. The agencies' use of the second performance criterion, relevance, is somewhat less apparent. The panel found that agencies recognize the importance of relevance in planning and review and that they consider the degree to which research programs and projects support their missions. However, although the use of relevance as an evaluation criterion is commonly embedded as an implicit element of planning and reviewing, it might not appear as an explicit element of published GPRA performance plans or reviews.

In addition, according to statements by agency officials, relevance appears to be evaluated at different stages by different people, most often by administrators who judge by custom or instinct whether a given line of research is relevant to a mission. Agencies' methods of performance review might therefore not be sufficiently clear to oversight groups and the public.

Recommendation M-3

Agencies should clarify their use of relevance as a criterion in evaluating their research programs. User groups should be part of the relevance evaluation process, and their role should be described clearly in performance plans and reports. Although agencies commonly use the criterion of relevance in implicit fashion, it should be made more visible to user groups, oversight bodies, and the public. Clear judgments about relevance can help agencies establish priorities among competing programs of equal scientific interest.

2.2.3 Leadership. COSEPUP indicated in its *Goals* report that US-supported research programs should be at least "among the leaders" in all major fields and that international benchmarking can provide a reasonably quick and inexpensive method of assessing the nation's leadership level. In general, however, agencies have not used the method of international benchmarking to evaluate the leadership level of research programs against world standards.

Most agencies are aware of the testing of international benchmarking by COSEPUP, and several agencies are considering its use. One impediment is that implementation would require additional time and resources. Agencies have used various other measures of leadership—such as international prizes, patents, national awards, and the judgment of experts—but not in a broad or standardized way.

In keeping with their diversity, agencies should devise their own approaches to evaluating leadership. They must first decide, for example, whether a particular field is one in which this country should be preeminent or simply among the leaders. They might also benefit from the use of existing expert review panels or other methods to evaluate leadership and from including international members as appropriate.

Recommendation M-4

Agencies should use international benchmarking to evaluate the leadership level of research programs, as described in COSEPUP'S earlier *Goals* and *International Benchmarking* reports, especially for emerging fields of research and those of national importance.
Agencies should select the fields to be evaluated and devise their own methods. If an agency does not evaluate a particular program with the criterion of leadership status, it should explain the reason for supporting the program (for example, a given program might have value for training or for filling gaps in knowledge important to the agency's mission).

2.3 Human Resources

Agencies justifiably attach great importance to their role in promoting the development of human resources. Their research programs depend on a continuing flow of talented scientists and engineers, who are best educated in the context of the research supported by agencies and other funders. However, this objective might not receive explicit emphasis or visibility in GPRA plans and reports.

Because the objective of developing human resources is generally not a clear or prominent feature of performance plans or reports, there is a risk of overlooking its continuing and fundamental importance especially in relation to the scientific and engineering research that is supported at universities. This objective must be explicit not only because it affirms the value of educating scientists and engineers by including them in the research programs of their advisers, but also because it demonstrates how reductions in research funding in specific fields can jeopardize the preparation of the next generation of scientists and engineers who will be important to the nation's future.

Recommendation M-5

The development of human resources should be emphasized as an explicit objective of GPRA performance plans and reviews. Plans to increase or reduce budgets should be described in terms of their impact on the future science and engineering workforce.

2.4 Aggregation

One aspect of GPRA that requires closer consultation between agencies and oversight groups is the clause that permits agencies to "aggregate, disaggregate, or consolidate program activities" in formulating GPRA plans and reports. Some difficulties appear to arise because of the different importance of research to various agencies. The portion of the budget allocated to research activities ranges from a small fraction (as in DOD and DOE) to most of or nearly all an agency's budget (as in NSF and NIH). Accordingly, agencies vary widely in the degree to which they have chosen to aggregate research programs for GPRA reporting. Some agencies report on individual programs; others describe entire research fields on an agency-wide basis.

A concern was voiced by representatives of several agencies in which research is a minor portion of the overall mission portfolio. The research divisions of such agencies might find it difficult to distinguish themselves from the dominant mission activities. Because the dominant activities tend to be easier to express in terms of predictable targets and quantifiable progress (for example, constructing a building, setting up a new social service, or planning a space launch), the agency's GPRA performance plans and reports are expressed primarily in terms of quantitative goals and milestones. Research programs in these agencies might find themselves compelled to conform to a prescribed reporting format.

When the degree of aggregation is high, oversight bodies, potential users, and the public might not be able to see or understand the detailed layers of decision-making and management that underlie the GPRA descriptions. In some instances, concerns were expressed by those advising oversight bodies that a high level of aggregation makes the underlying processes less clear. Agencies indicated that they choose a high degree of aggregation because individual program activities are not easily linked to budgetary line items or because specific mechanisms of decision are too numerous to discern at the high aggregation level of GPRA reporting. Because a primary purpose of GPRA is to permit oversight bodies to understand how agencies make decisions and set priorities, it is essential that these bodies be able to see the connections between performance plans, performance reports, and strategic plans.

Recommendation M-6

Agencies that choose to aggregate their research-program activities at a high level should endeavor to make clear the decision-making processes that lie below this level. A degree of transparency is needed for oversight bodies and the public to understand how an agency evaluates its programs and sets priorities. Although oversight bodies cannot review the thousands of subentities that perform their own planning and reviewing within agencies, they can reasonably expect access to documents that help them answer specific questions.

2.5 Validation and Verification

Although expert review has long been the accepted method of evaluating research in science and engineering, some aspects of the implementation of expert review are unclear to outside observers. Oversight bodies and some agencies express a need for clearer validation and verification of expert review, such as explication of how agencies establish the independence of reviewers. For their part, agencies do not customarily communicate the details of how they validate or verify their evaluation procedures in ways that are clear to oversight groups or users. Validation is particularly of concern when the level of aggregation is high (that is, where research fields are evaluated as a single program on an agency-wide basis). The process of expert review is implicitly understood by those involved in research, because agencies consider expert review to be the most objective and reliable mechanism for evaluating their research programs. However, the mechanism must be explicitly and publicly articulated to those who are charged with oversight and who might be less familiar with or accepting of the custom.

Recommendation M-7

Agencies should devise ways to describe how they validate their research-evaluation methods, describ-

ing, for example, how they select expert reviewers and choose to aggregate research programs for review.

2.6 Summary

The agencies examined have devoted considerable effort to developing reporting procedures that comply with the requirements of GPRA and are congruent with their internal planning procedures. However, some expressed a need for new processes. It was clear from the panel's discussions with agencies that compliance methods are still very much "works in progress" and that further work is needed if agencies are to both fulfill the intent of the law and provide benefits to the agencies.

Testimony during the focus groups indicated that the three criteria (especially quality and relevance), as described by COSEPUP, had proved useful in approaching the requirements of GPRA. In particular, the panel was able to verify the usefulness of the criteria to the agencies themselves. The panel concluded that the criteria of quality, relevance, and leadership are more effective than quantitative performance indicators for evaluating research programs.

CHAPTER 3

COMMUNICATION ISSUES

One of the challenges to agencies and oversight bodies is to interpret the law itself in ways that achieve the desired results of accurate reporting and accountability. The language of the law is general; it addresses the agencies as a single population without distinguishing among them. As mentioned above, however, the variations in structure and function of agencies are considerable. For the most part, individual agencies have been left with the task of working out for themselves the best way to interpret such terms as "program activity," "performance indicator," and "program result" within their particular structures.

In the absence of detailed, continuing discussions among the creators of the law, oversight bodies, and agencies, the agencies have little guidance on the best ways to apply such terms to existing agency procedures and research programs. In addition to its comments on these communication difficulties, the panel offers several specific observations about the issue of oversight.

The issue of communication is fundamental to the implementation of GPRA. That is, the essential objective of the law, in the context of research, is to improve the management of government services. And a key element of that management is communication of federal agencies with oversight bodies about the scientific and energy research needs of the nation as a whole. To achieve that objective, long-standing terms and customs of the agencies must be translated and communicated so that they are clear outside the agencies. If this can be achieved, internal "language" and GPRA "language" can be reconciled toward the dual goals of facilitating congressional budgeting activities and enhancing the effectiveness and efficiency of agency management.

This chapter examines the communication between the agencies and the primary audiences for its GPRA reports: oversight groups, the users of research, and the public. As was the case with the previous chapter, the observations here are based on the panel's focus groups and workshop where agency and oversight group representatives discussed agency responses to the following questions:

- How is the result communicated to different audiences (e.g., S&T community, advisory committees, agency leadership, administration, Congress)?
- How is the result used in internal and external decision-making?

The agency responses are summarized in Appendix C and the workshop discussion is provided in Appendix D.

3.1 Communication Between Agencies and Oversight Groups

The viewpoints of Congress, GAO, OMB, and other entities interested in the implementation of GPRA vary with their specific charges. In general, however, all of them have expressed a desire to know more about:

- What procedures the federal agencies use to comply with GPRA;
- How successful those procedures are; and
- How the GPRA planning and reporting processes can serve agency missions and the public interest better than is available in the existing documentation.

The panel's discussions with agency and oversight representatives made it clear that communication between agencies and oversight groups must be more regular, extensive, and collaborative to facilitate agency responses to GPRA. Participants in the focus groups and workshop emphasized that improved communication about methods could hasten the implementation of GPRA, increase its value as a planning and accountability tool, and reduce the cost of compliance.

One common complaint from agencies is that oversight bodies are quicker to criticize shortcomings than to suggest improvements or specify desired outcomes. The most constructive course would be for oversight bodies to suggest how agencies can use procedures already in place without adding additional steps. GPRA should not "make extra work," and oversight bodies should be willing to work with agencies to ensure that this does not happen. The public benefit of such a course is to eliminate unnecessary cost and duplication of effort.

As mentioned above, one objective of GPRA and of the oversight bodies is to clarify the mechanisms used by agencies to validate and verify their evaluation procedures. For example, oversight representatives would like to be assured that the reviewers of research programs are objective, experienced, and expert—again, a communication issue. Oversight bodies have expressed an inability to see or understand how those qualities are validated by agencies, and they have asked for improved communication about the procedures.

In addition, agencies sometimes receive conflicting signals from oversight groups. Even from a single oversight entity, they might receive different guidance from different staff members. One agency revised its accounting to link GPRA reports more tightly with its budgets, for example, only to be told by a congressional oversight committee to return to the previous format to which the committee was accustomed.

Further, different congressional committees prefer different levels of aggregation. Some prefer a high level, in which many or all research programs are considered together; others prefer a disaggregated approach. This presents a confusing picture to agencies.

Most importantly, agencies have not yet seen the use of their reports in the congressional decision-making that determines the size and form of their budgets. That could reduce the incentive of agencies to integrate their own planning and budgeting functions with the requirements of GPRA. Without such integration, agencies duplicate their reporting efforts to serve internal budgeting functions and GPRA requirements.

Recommendation C-1

Agencies and oversight groups should strive to communicate more effectively with each other so as to improve agencies' progress in implementing GPRA.

During focus groups, the workshop, and interviews, it was consistently clear that improved communication between these two sectors could reduce the difficulties and misunderstandings experienced by some agencies. Agencies should provide brief, clear summaries of the procedures by which they perform expert review, aggregate programs, validate evaluation methods, set research priorities, and include user groups and other members of the public in planning and reporting. That simple step would allow a clearer view of the links between GPRA documents and agencies' internal procedures.

3.2 Communication by Agencies with User Groups and the Public

User groups, as described above, represent important segments of the public that are served by publicly funded research. They have important roles to play in planning and evaluating research programs. Some agencies make good use of such groups on panels and review committees, but their activities are seldom made clear in GPRA documents or to oversight groups. Agencies can demonstrate the value and operation of their review processes better by publicly describing them to oversight groups, the potential users of research results, and the general public.

Recommendation C-2
Agencies should seek to demonstrate more clearly to users and the public how they set priorities for evaluating research programs.

3.3 Communication by Oversight Groups

Good oversight need not impede agencies that are making a sincere effort at compliance. Oversight activity is likely to be lowest for agencies that have advanced in their efforts to comply with GPRA and greatest for those still struggling. Agencies, for their part, should strive to make the research clear to nonscientists among oversight bodies.

Recommendation C-3
Oversight groups should provide more clarity and consistency in their expectations of agencies that are striving to comply with the requirements of GPRA. They should consult, as requested, on practical ways to integrate agencies' internal planning and reviewing practices with GPRA requirements. They should also meet more often among themselves to coordinate their expectations of agency practices.

3.4 Communication Within Agencies

One objective of the law is to encourage the integration of program activities and strategic planning. With or without GPRA, in fact, each agency can benefit from reviewing its research programs in the light of how they and their individual projects serve the broader strategic plan. The strategic plan, in turn, should evolve year by year in view of the changes that are made in individual programs and projects. More effective use of strategic planning can allow oversight groups to understand the contribution of individual programs to an agency's mission and hence improve agency-oversight communication.

Recommendation C-4
Agencies, especially large mission agencies, should seek to improve internal communication about GPRA so that the evaluation of research activities is not hidden within the agency's overall GPRA reporting.

3.5 The Issue of Timing

Although agency representatives expressed enthusiasm for using the criterion of quality to evaluate research on a regular basis, they voiced concern over the requirement to provide annual reports for their basic research programs. As explained earlier, basic research often does not produce useful results in a single year and must be monitored over several years before outcomes become apparent. A particular concern on the part of agency representatives is that programs and individual researchers might feel pressured to produce evidence of annual achievement in the form of "extra" publications or other meaningless metrics. Such activities would waste valuable resources and distract researchers from productive work.

Because the value of investments in basic research can be evaluated only over long periods, retrospective methods might be more effective than annual reports. For example, NSF is experimenting with "rolling" assessments whereby one-third of the portfolio is evaluated each year. Every research project is thus evaluated every 3 years, a reasonable period in which to expect results.

NSF suggests that that method could be useful for other agencies and that the 3-year focus be applied to performance plans, as well as performance reports. Thus, an agency would set 3-year performance targets, rather than annual performance targets, on research goals. Performance reports would still be annual, but they would cover one-third of the portfolio each year. They would also describe trends in the direction of basic research, the rate of progress of research, and the productivity of special initiatives. Management goals and short-term objectives in applied-research programs, where targets are more easily calibrated and predicted, would still be described annually in performance plans and reports.

A potential benefit of GPRA is the ability to strengthen agencies' planning procedures by making available the research results of previous years. Because of timing requirements built into the legal guidelines of GPRA, agencies find that they must begin work on future performance plans before the most recent performance reports are available. For example, in November 2000, one agency was beginning its performance plan for 2002 before it had finished its performance report for 2000.

One reason for the difficult timing is that the act was designed to enable oversight groups to connect each performance plan and performance report directly with its corresponding annual budget. The timing is unfortunate for several reasons. Agencies and researchers need the flexibility to change the course of a research project if change is warranted by previous and current results. And neither agencies nor the public receive a benefit when agencies create detailed performance plans before they have sufficient recent information on the performance of current programs.

Recommendation C-5

Agencies should work with oversight bodies to create more-realistic GPRA reporting schedules. Such schedules should recognize the important difference between research programs of differing goals and time

frames. Although yearly reporting may be appropriate for applied research, a 3-year (or longer) performance schedule for basic research would usually be more suitable and valuable. The schedules should allow agencies to use previous results when preparing performance plans. Agencies should also continue their efforts to integrate GPRA planning and evaluation procedures into current agency processes.

Specifically, the panel suggests that agencies engaged in basic research make 3-year performance plans and set 3-year performance targets for research goals in their performance plans, rather than targets that refer to particular fiscal years. Management goals and short-term objectives in applied-research programs should still refer to a 1-year period. Performance reports should be annual and stress trends and indicators of the direction of basic research and the level of progress and productivity of special initiatives.

3.6 Summary

Communication between agencies and oversight bodies is essential to making the GPRA process work. So far, the communication process has been flawed from the viewpoint of both sides. If agencies are clearer regarding their methodology and oversight groups are clearer and more consistent regarding their expectations, a better and more useful product will result.

CHAPTER 4

CONCLUSIONS AND RECOMMENDATIONS

Over the last 4 years, federal agencies that support research in science and engineering have moved by stages toward full implementation of GPRA. The central objective of the act is to elicit from the agencies a regular accounting of the planning, performance, and results of their research activities.

Agencies have spent substantial time and effort in devising ways to implement the act. However, both the agencies and oversight bodies must still develop better refinements to improve interpreting, implementing, and communicating with each other about GPRA.

To assist in the complex processes of implementing GPRA, this report has attempted to summarize and interpret the experiences of agencies and oversight bodies. In particular, its major sections examine the current process and recommend the most appropriate methods of evaluating basic– and applied–research programs, the criteria that agencies can and should use to perform their evaluations, and the experiments and difficulties experienced by agencies in communicating their evaluation results internally and externally.

After its study of GPRA with agency and oversight personnel, the present panel has concluded that the manner of planning and evaluating research programs carries great importance. It is apparent that inappropriate methods and inadequate communication can harm the programs that the law seeks to strengthen. We hope that the general observations, conclusions, and recommendations in this report help agencies and oversight groups as they continue to take the incremental steps necessary to implement GPRA for the country's federal research programs.

Chapters 2 and 3 each contain specific recommendations for agencies and oversight bodies that are designed to expedite the implementation of GPRA. This chapter offers a brief set of more-general conclusions and recommendations that consolidate the major themes of the preceding text.

4.1 General Conclusions

The panel offers the following 10 conclusions:

Conclusion 1: All five agencies have made a good-faith effort to develop reporting procedures that comply with the requirements of GPRA. Some agencies stated that GPRA compliance has added substantially to the cost of their planning and evaluation activities in the form of staff time and resources. Others report that they have been able to integrate GPRA with their traditional budget and planning processes although at some cost of time and effort.

Conclusion 2: Some agencies are using the GPRA process to improve their operations. These agencies report benefits in strengthening program management and enhancing communication about their programs to the users of research and the general public. The need to do so depends on the goal of that agency and the degree to which there is concern about a given field of research or about new and emerging programs.

In promoting greater accountability, the act calls for firmer alignment of research programs with overall strategic planning and for a higher degree of accountability. These agencies report progress on both counts—in strengthening the management of their programs and in enhancing their ability to communicate the value of their programs to the users of research and the public.

However, while some agencies report that they have been

able to derive their GPRA requirements from the same management processes that they traditionally use for internal control and budgeting, others see GPRA requirements as extra burdens that add to the planning and reporting workload, with lost opportunities in terms of costs of staff time and resources devoted to this requirement.

Conclusion 3: The most effective technique for evaluating research programs is review by panels of experts using the criteria of quality, relevance, and, when appropriate, leadership. Agency approaches to GPRA research programs demonstrate the utility of expert review using the same criteria of quality and relevance as outlined in COSEPUP's original report. The international leadership criteria is generally not evaluated by most federal agencies at this time, although several are interested in such a measure. However, given the diversity in mission, complexity, culture, and structure of federal agencies that support research, it is not surprising that their approaches to GPRA have varied. One size definitely does not fit all.

Conclusion 4: Oversight bodies and some agencies need clearer procedures to validate and verify agency evaluations. In particular, oversight bodies expressed a desire for better understanding of the methodology and results of expert review evaluations.

Conclusion 5: Agencies choose to aggregate their research programs at different levels. Some agencies provide evaluations on a field-specific or program-specific basis; others do so for the research program in its entirety. Aggregating at a high level can make it difficult for oversight bodies to clearly see and understand the methods and programs that are the focus of the analyses.

Conclusion 6: The development of human resources as an agency objective sometimes does not receive explicit emphasis or visibility in GPRA plans and reports. When this objective is explicit, it not only affirms the value of the US tradition that includes graduate students in the research programs of their advisers—but also shows how reductions in research funding can jeopardize the preparation of the scientists and engineers the nation will need in the future.

Conclusion 7: Agencies often receive conflicting messages from oversight bodies about the desired format, content, and procedures to be used in GPRA compliance. For example, one agency made an effort to tie its GPRA reports more closely to its annual budget, as required in the act, only to be told by a congressional committee to return to a previously used format—another was told the reverse.

Conclusion 8: Due to timing requirements built into the legal guidelines of GPRA, agencies find that they must begin work on performance plans before the relevant performance reports are complete. As a result, the potential benefit of GPRA in providing a mechanism for incorporating performance results of previous years into performance plans for later years is limited.

Conclusion 9: Communication between agencies and oversight groups is not sufficiently regular, extensive, or collaborative. During focus groups, the workshop, and interviews, it was consistently clear that improved communication between these two sectors could reduce the difficulties and misunderstandings experienced by some agencies.

Conclusion 10: The degree to which the results of GPRA results of research programs are being used by

oversight groups for programmatic decision-making are uncertain. Are the results of the "results act" being used? In particular, agencies have not yet seen the use of their reports in the congressional decision-making that determines the size and priorities of their budgets.

4.2 General Recommendations

On the basis of these observations, the panel offers the following general recommendations:

Recommendation 1: Federally supported programs of basic and applied research should be evaluated regularly through expert review, using the performance indicators of quality, relevance, and, where appropriate, leadership.

The language of the act strongly urges agencies to evaluate their programs annually through the use of quantitative measures so that progress can be followed with clear numerical indicators. The panel reaffirms COSEPUP's earlier assertion that research programs, especially those supporting basic research, cannot be meaningfully evaluated this way annually. Instead, these programs can be evaluated over a somewhat longer term through expert review, which has a long tradition of effectiveness and objectivity.

Recommendation 2: Agencies should continue to improve their methods of GPRA compliance and to work toward the goals of greater transparency, more-realistic reporting schedules, clear validation and verification of methods, and the explicit use of the development of human resources as an indicator in performance plans and reports.

Transparency refers to the ability to readily see how and why an agency decides to emphasize or de-emphasize a particular program or area of research. When an agency describes its performance plans and reports from an agencywide point of view (for example, an agency might describe its efforts to reduce global warming as though it were a single program), it is difficult for oversight bodies or the public to understand the process of priority-setting. Although oversight bodies or agents of the public would not be expected to review the thousands of subentities that perform their own planning and reviewing within agencies, they can reasonably expect access to documents that help them to answer specific questions.

Although GPRA requires annual reporting on all programs, basic research often does not produce useful results in a single year and must be monitored over several years before outcomes become apparent. Agencies should experiment with alternative reporting forms, as permitted by GPRA, that provide realistic evaluations of long-term research.

Although expert review has long been the accepted method for evaluating research in the science and engineering communities, some aspects of its performance are unclear to outside observers. Agencies should make clear how they validate their research-evaluation methods, such as the means by which they select expert reviewers and choose to aggregate research programs for review.

Agencies have a large stake in the education and training of scientists and engineers, but this objective might not receive explicit emphasis or visibility in GPRA plans and reports. The objective must be explicit not only because it affirms the value of educating young scientists and engineers in the context of research, but also because it demonstrates how reductions in research funding could weaken the corps of human resources that are essential for the nation's future.

Recommendation 3: Agencies and oversight bodies should work together as needed to facilitate agencies integrating their GPRA requirements with their internal

planning, budgeting, and reporting processes. In addition, they should work together to adjust the timing of GPRA reporting to capitalize on the value of the planning process.

Whenever possible, agencies should use procedures already in place without adding steps. GPRA should not add unnecessarily to the workload of agencies, and oversight bodies should help agencies to ensure that this does not happen. At the same time, effective linkage of GPRA reporting with budgets may help agencies explain their needs to Congress and justify funding levels during periods of restrained budgets.

Recommendation 4: Agencies should strive for effective communication with oversight groups on the implementation of GPRA. For their part, oversight bodies should clarify their expectations and meet more often among themselves to coordinate their messages to agencies.

A principal purpose of GPRA is to improve how agencies communicate their results to oversight groups, the "users" of research, and the general public. More-effective communication will enhance the value of the act to all constituents.

As indicated in COSEPUP's first report, GPRA is potentially useful because it "provides an opportunity for the research community to ensure the effective use of the nation's research resources in meeting national needs and to articulate to policymakers and the public the rationale for and results of research." However, the act will not fulfill its intended objectives unless the Senate and House Operations committees, working with OMB, identify and respond to agency concerns through open discussion. Unless the agency responses to GPRA are useful to Congress in the urgent task of setting priorities and budgeting, the value of the act might not warrant the time and effort it requires of the federal government.

4.3 Specific Recommendations

Provided below are the specific recommendations that are scattered throughout this report:

4.3.1 Agency Methods

Recommendation M-1: Agencies should continue to take advantage of their existing expert review panels, but should review the balance in their membership, particularly the need to include user groups, and the time panel members devote to GPRA versus other topics so that it is not excessive. In addition, they should review the degree to which internal vs. external reviewers are used.

Recommendation M-2: Agencies should continue to use peer review to evaluate the quality of their research programs.

Recommendation M-3: Agencies should clarify their use of relevance as a criterion in evaluating their research programs. User groups should be a part of the relevance evaluation process, and their role should be described clearly in performance plans and reports.

Recommendation M-4: Agencies should use international benchmarking to evaluate the leadership level of research programs, as described in COSEPUP's earlier *Goals and International Benchmarking* reports, especially for emerging fields of research and those of national importance.

Recommendation M-5: The development of human resources should be emphasized as an explicit objective of GPRA performance plans and reviews.

Recommendation M-6: Agencies that choose to aggregate their research-program activities at a high level should endeavor to make clear the decision-making processes that lie below this level.

Recommendation M-7: Agencies should devise ways to describe how they validate their research-evaluation methods, describing, for example, how they select expert reviewers and choose to aggregate research programs for review.

4.3.2 Communication

Recommendation C-1: Agencies and oversight groups should strive to communicate more effectively with each other so as to improve agencies' progress in implementing GPRA.

Recommendation C-2: Agencies should seek to demonstrate more clearly to users and the public how they prioritize and evaluate research programs.

Recommendation C-3: Oversight groups should provide more clarity and consistency in their expectations of agencies that are striving to comply with the requirements of GPRA.

Recommendation C-4: Agencies, especially large mission agencies, should seek to improve internal communication about GPRA so that the evaluation of research activities is not hidden within the agency's overall GPRA reporting.

Recommendation C-5: Agencies should work with oversight bodies to create more-realistic GPRA report-

ing schedules. Such schedules should recognize the important difference between research programs of differing goals and time frames. While yearly reporting may be appropriate for applied research, a 3-year (or longer) performance schedule for basic research would usually be more suitable and valuable.

4.4 Summary

Much has been learned about the procedures of planning, evaluation, and management in the last several years, and some value will have been gained by the agencies from their own discussion of accountability. However, one key remaining question is the degree to which oversight groups are using the results of the "results act" for programmatic decision-making. Unless the agency responses to GPRA are useful to Congress in the urgent task of setting priorities and budgeting, the value of the act might not warrant the time and effort it requires of the federal government. But by working more closely together than they have in the past, the federal agencies and the oversight bodies can implement the letter and spirit of GPRA in ways that lead to greater efficiency, lower cost, and more effective research programs that are demonstrably conducted in the national interest.

Appendixes

PANEL AND STAFF BIOGRAPHICAL INFORMATION

Enriqueta C. Bond (Cochair) received her undergraduate degree in zoology and physiology from Wellesley College, a master's degree in biology and genetics from the University of Virginia, and a PhD in molecular biology and biochemical genetics from Georgetown University. She is a member of the American Association for the Advancement of Science, the American Society for Microbiology, the American Public Health Association, and the Institute of Medicine (IOM). Dr. Bond was IOM's executive officer from 1989 to 1994. She became President of the Burroughs Wellcome Fund in July 1994. She serves on the IOM Council, chairs the Board of the North Carolina Biotechnology Center, the Board of Scientific Counselors of the National Centers for Infectious Diseases, and co-chairs the IOM Clinical Research Roundtable.

Alan Schriesheim (Cochair) is director emeritus of Argonne National Laboratory (ANL). He also served as senior department director, COO, director, and CEO of ANL. Previously, he worked at Exxon Research and Engineering Co. in a variety of positions, including general manager of the Technology Department, assistant manager, assistant director, and director of corporate research laboratories. Dr. Schriesheim has been honored with the Award in Petroleum Chemistry by the American Chemical Society and the Karcher Silver Medallist Lecturer. He has served on many national and international boards and committees. Most recently, he served as a member of the National Academy of Engineering's selection committee for determining the Greatest Engineering Achievements of the 20th century. Dr. Schriesheim is a member of the National Academy of Engineering. He is a fellow of the New York Academy of Sciences and the American Institute of Chemists and a member of the American Association for the Advancement of Science, the American Chemical Society, and Sigma Xi. Dr. Schriesheim received his PhD in physical organic chemistry from Pennsylvania State University.

John E. Halver is professor emeritus in nutrition at the School of Aquatic and Fishery Sciences at the University of Washington. He was a laboratory director at the US Fish and Wildlife Service for 25 years and senior scientist for the US Department of Interior (1975-1978). Concurrently, he served as a research officer in the US Army Medical Research and Nutrition Laboratories for 30 years. As president of Halver Corporation, an ecosystem management consulting company, he has been retained by the World Bank, the United Nations Development Program, the Food and Agriculture Organization of the United Nations, the US Agency for International Development, and several other agencies for technology-transfer projects in over 30 countries. He has published over 185 peer-reviewed articles and edited seven books during his research activities (1950-2000) and has been chair or member of several National Research Council committees. His professional memberships include the American Institute of Fisheries Research Biologists (Fellow), the American Institute of Nutrition (Fellow), the American Fisheries Society (Certified Scientist), Society of Experimental Biology and Medicine, World Aquaculture Society, and the American Chemical Society (Senior Grade). Dr. Halver was elected to the National Academy of Sciences in 1978, the Hungarian Academy of Sciences in 1998, and the National Fish Culture Hall of Fame in 2000. He received his PhD in biochemistry from the University of Washington.

Brigid L.M. Hogan is a Howard Hughes Medical Institute investigator and Hortense B. Ingram Professor in the Department of Cell Biology at Vanderbilt University School of Medicine. Before joining Vanderbilt, she was head of the Laboratory of Molecular Embryology, first at the Imperial Cancer Research Fund and then at the National Institute of Medical Research in London. Dr. Hogan is a member of the National Academies Committee on Science, Engineering, and Public Policy. She is also a member of the Institute of Medicine and the European Molecular Biology Organization. Dr. Hogan received her PhD in biochemistry from Cambridge University, England.

Wesley T. Huntress, Jr. is the director of the Carnegie Institution's Geophysical Laboratory. He was associate administrator for space science at National Aeronautics and Space Administration Headquarters from 1993 to 1998 and director of the Solar System Exploration Division from 1990 to 1992. Before joining the Senior Executive Service, Dr. Huntress had been detailed from the California Institute of Technology's Jet Propulsion Laboratory (JPL) for 2 years as special assistant to the director of the Earth Science and Applications Division. Dr. Huntress began his career at JPL in 1968, first as a National Research Council resident associate before joining JPL permanently. Dr. Huntress has over 100 peer-reviewed publications in astrochemistry. He is a member of the NRC Division on Engineering and Physical Sciences. His current professional memberships include the American Astronautical Society (past President), American Astronomical Society Division of Planetary Sciences (Vice Chair), and Vice President of the Planetary Society. Dr. Huntress received his PhD in chemical physics from Stanford University.

Louis J. Lanzerotti is Distinguished Member of the technical staff of Bell Laboratories, Lucent Technologies, where his research interests have included geophysics and space plasma physics as related to planetary magnetospheres and atmospheres, energetic particles emitted by the sun, and the engineering impacts of natural and artificial space phenomena on space and terrestrial technologies. Dr. Lanzerotti is a member of the National Academy of Engineering and the International Academy of Astronautics, the author or co-author of more than 500 papers, and co-author of three books. He is a fellow of the Institute of Electrical and Electronics Engineers, the American Institute of Aeronautics and Astronautics, the American Geophysical Union, and the American Physical Society. He is a member of the Governing Board of the American Institute of Physics. He received the National Aeronautics and Space Administration Distinguished Public Service Medal in 1988 and 1994 and has both an Antarctic mountain and a "minor planet" named for him. Dr. Lanzerotti received his PhD in physics from Harvard University.

Rudolph A. Marcus is Arthur Amos Noyes Professor of Chemistry at California Institute of Technology. He has been a member of several National Academies committees. Numerous awards—including the Nobel Prize in chemistry in 1992, the Wolf Prize in 1985, and the National Medal of Science in 1989—distinguish Dr. Marcus's career, in addition to various honorary doctorates and professorships. Dr. Marcus is a member of the National Academy of Sciences. He is a foreign or honorary member of many societies, including the American Chemical Society, the American Physical Society, the American Academy of Arts and Sciences, the American Philosophical Society, the Royal Society of London, the Royal Society of Canada, and the Chinese Academy of Sciences. Dr. Marcus received his PhD in physical chemistry from McGill University.

Stuart A. Rice is the Frank P. Hixon Distinguished Service Professor at the University of Chicago's James Franck Institute. His research interests include elementary photophysical and photochemical processes, quantum chaos, active control of selectivity of chemical reaction, reaction-rate theory, study of liquid surfaces, properties of supported monolayers, phase transitions in interfaces, and the equilibrium and dynamical properties of quasi-two-dimensional systems. Dr. Rice is a member of the National Academy of Sciences. He has served on many national and international boards and committees as editor, member, and consultant, including the National Science Board and National Academies' Board on Chemical Sciences and Technology. Dr. Rice is a 2000 recipient of the National Medal of Science. He received his PhD in chemistry from Harvard University.

Herbert H. Richardson is Regents Professor and Distinguished Professor of Engineering, associate vice chancellor for engineering, and director of the Texas Transportation Institute at the Texas A&M University System. Before taking his position at Texas A&M, he was chief scientist at the US Department of Transportation and professor and associate dean of engineering at MIT. He has chaired and served on numerous National Academies committees examining transportation issues. Dr. Richardson is a member of the National Academy of Engineering. His professional memberships include the American Society of Engineering Educators, the New York Academy of Science, the American Association for the Advancement of Science, the American Society of Mechanical Engineers, Sigma Xi, and the Institute of Electrical and Electronics Engineers. He received his ScD in mechanical engineering from Massachusetts Institute of Technology.

Max D. Summers is a Distinguished Professor at Texas A&M with joint appointments in biology, biochemistry and biophysics, entomology and genetics. He is holder of the Chair in Agricultural Biotechnology; Houston Intellectual Property Association, 1999 Outstanding Inventor of the Year. He is the director of the Center for Advanced Invertebrate Molecular Sciences. Dr. Summers is a member of the National Academy of Sciences and has participated in many National Academies activities. He is a Fellow of the American Academy of Microbiology. His professional memberships also include the American Society for Microbiology, American Society for Virology (past president), American Association for the Advancement of Science, American Society for Biochemistry and Molecular Biology and the American Society for Cell Biology. Dr. Summers received his PhD in entomology from Purdue University.

Morris Tanenbaum was the vice chairman of the Board and chief financial officer of AT&T from 1988 to 1991. He began his career at Bell Laboratories on the technical staff and held various positions at Western Electric Company, including vice president of the Engineering Division and vice president of Manufacturing, before returning to Bell Laboratories in 1975 as executive vice president. In 1978, he became president of New Jersey Bell Telephone Company; he returned to AT&T as executive vice president for corporate affairs and planning in 1980, becoming the first chairman and CEO of AT&T Communications in 1984. Dr. Tanenbaum is a member of the National Academy of Engineering. He has chaired and served on numerous National Academies committees. He is a fellow at the American Academy of Arts and Sciences, the American Physical Society, and the Institute of Electrical and Electronic Engineers, and a member of the American Chemical Society and the American Institute of Mining, Metallurgical, and Petroleum Engineers. Dr. Tanenbaum received his PhD in physical chemistry from Princeton University.

Bailus Walker, Jr. is professor of environmental and occupational medicine at Howard University. Before his position at Howard, he was dean of the College of Public Health at the University of Oklahoma Health Science Center. Dr. Walker is a member of the Institute of Medicine. He has served on many National Academies committees and is a Fellow of the Royal

Society of Health and Distinguished Fellow of the American College of Epidemiology. Dr. Walker received his PhD in occupational and environmental health from the University of Minnesota.

Robert M. White is University Professor of Electrical and Computer Engineering and director of the Data Storage Systems Center at the Carnegie Mellon University (CMU). His interests encompass technology and technology-policy issues. His policy interests are focused on federal science and technology policy. Before joining CMU, he served during the George H.W. Bush administration as the first undersecretary of commerce for technology. Dr. White was a principal scientist at Xerox's Palo Alto Research Center and chief technical officer of Control Data Corporation. His early career was spent in teaching and research. He is a member of the National Academy of Engineering. He has participated in numerous activities of the National Academies. Dr. White's professional memberships include the American Physical Society, the Institute of Electrical and Electronics Engineers, and the American Association for the Advancement of Science. He received his PhD in physics from Stanford University.

Staff and Consultants:

Deborah D. Stine is associate director of the Committee on Science, Engineering, and Public Policy (COSEPUP). She has worked on various projects throughout the National Academies since 1989. She received a National Research Council group award for her first study for COSEPUP on policy implications of greenhouse warming and a Commission on Life Sciences staff citation for her work in risk assessment and management. Other studies have addressed graduate education, responsible conduct of research, careers in science and engineering, environmental remediation, the National Biological Survey, and corporate environmental stewardship. She holds a bachelor's degree in mechanical and environmental engineering from the University of California, Irvine; a master's degree in business administration; and a PhD in public administration, specializing in policy analysis, from the American University. Before coming to the National Academies, she was a mathematician for the US Air Force, an air-pollution engineer for the state of Texas, and an air-issues manager for the Chemical Manufacturers Association.

Alan H. Anderson is a consultant writer for the Committee on Science, Engineering, and Public Policy (COSEPUP), has worked on a number of recent reports, including *Science, Technology, and the Federal Government: National Goals for a New Era*; *Reshaping the Graduate Education of Scientists and Engineers*; *Capitalizing on Investments in Science and Technology*; *Evaluating Federal Research Programs: Research and the Government Performance and Results Act*; and guides for students and faculty on careers in science and engineering and on mentoring students in science and engineering. He also writes for the Institute for Advanced Study and other clients. He has been a science writer for *Time* magazine and other publications. He holds a master's degree from the Columbia University School of Journalism and a BA in English from Yale University.

Susan E. Cozzens is professor and chair of the School of Public Policy at the Georgia Institute of Technology. She spent 11 years on the faculty of Rensselaer Polytechnic Institute. From 1995 through 1997, Dr. Cozzens was director of the Office of Policy Support at the National Science Foundation (NSF). The office coordinated policy and management initiatives for the NSF director, primarily in peer review, strategic planning, and assessment. She has been an invited speaker, nationally and internationally, on science policy and research evaluation. She has served as a consultant to many organizations in both private and government sectors, including the National Academies Committee on Science, Engineering, and Public Policy; the Institute of Medicine; the Office of Science and Technology Policy; NSF; the General Accounting

Office; and the Department of Health and Human Services. Dr. Cozzens has a distinguished record of publication and service in science policy and science and technology studies. Her PhD is in sociology from Columbia University (1985), and her bachelor's degree from Michigan State University (1972, summa cum laude). She is a recipient of Rensselaer's Early Career Award, a member of Phi Beta Kappa and Phi Kappa Phi, and a fellow of the American Association for the Advancement of Science.

David M. Hart is associate professor of public policy, Kennedy School of Government, Harvard University. His research interests are at the intersection of American political development, political economy, and science and technology policy. His book, *Forged Consensus: Science, Technology, and Economic Policy in the United States, 1921-1953*, was published by Princeton University Press in 1998. His current research focuses on the politics of high-technology businesses and on new approaches to civilian technology policy. He has also written on the organization of the executive office of the president, genetic discrimination, environmental policy, and the "foreign policies" of US research universities. He serves as faculty chair for the political advocacy methodological area of concentration. He is a member of the Whitehead Institute Task Force on Genetic Testing, Privacy, and Public Policy; a member of the Academic Advisory Board, Center on Science, Policy and Outcomes; and a member of the Master's in Public Policy Admissions Committee. His PhD is in political science from the Massachusetts Institute of Technology, and his bachelor's degree is from Wesleyan University.

APPENDIX B

WHITE HOUSE AND CONGRESSIONAL CORRESPONDENCE

THE WHITE HOUSE

WASHINGTON

April 6, 1999

Dr. Bruce Alberts
President
National Academy of Sciences
2101 Constitution Avenue, NW
Washington, DC 20418

Dear Bruce:

The VA-HUD legislation, which passed last session in the omnibus appropriation bill, called on the Office of Science and Technology Policy (OSTP), the Office of Management and Budget (OMB), and the National Academy of Sciences to consider the need for a study on "Accountability of Federally-Funded Research." As discussed at our recent meeting, I believe the Academy could be useful in assisting the agencies' efforts to craft GPRA plans and reports that are responsive to the law, OMB guidance, and their missions.

I would like for the Academy to undertake an independent assessment of the strategic and performance plans the agencies have developed and of the responsiveness of their performance reports, which are due in March 2000. As you are aware, we and OMB view the implementation of the Government Performance and Results Act as a "work in progress," and we envision that agencies will benefit by sharing best practices in their performance reports, as they did in sharing their performance plans. This assessment should take into account the agencies' missions and how science and technology programs and human resource needs are factored into their GPRA plans.

Rather than assess GPRA documents for all federal S&T agencies, I believe a case studies approach on major programs of five or so different agencies should provide adequate coverage. Such a study would be timely for all federal agencies as they enter the next phase of developing their new strategic and performance plans as well as their subsequent performance report for the year 2000. It would also provide an opportunity for the Academy to suggest specific applications of recommendations from its earlier GPRA report.

In conducting this study, we hope you take the opportunity to hear from the various stakeholders in the process and work with the research agencies. In particular, the workshops COSEPUP held throughout its GPRA study proved useful in facilitating dialogue among the executive branch agencies, researchers, and Congress on key issues. If possible, they should be a key part of this study as well.

We look forward to reviewing the results of this effort.

Sincerely

Neal

Neal Lane
Assistant to the President
for Science and Technology

U.S. HOUSE OF REPRESENTATIVES

COMMITTEE ON SCIENCE

SUITE 2320 RAYBURN HOUSE OFFICE BUILDING
WASHINGTON, DC 20515–6301
(202) 225–6371
TTY: (202) 226–4410
http://www.house.gov/science/welcome.htm

May 12, 1999

Dr. Bruce Alberts
President
National Academy of Sciences
2 1 01 Constitution Avenue, NW
Washington, DC 20418

Dear Dr. Alberts:

Thank you for your recent report "Evaluating Federal Research Programs: Research and the Government Performance and Results Act." As you know, we issued a press release supporting the findings of this report and indicating a desire to turn those principles into practical processes at agencies.

After reviewing the April 6, 1999 letter from Dr. Neal Lane, Director of OSTP, we believe the proposed "accountability study" will provide a valuable service to the research community. In that letter, Dr. Lane requests the Academy "undertake an independent assessment of the strategic and performance plans the agencies have developed and of the responsiveness of their performance reports" to the Government Performance and Results Act (Results Act) via a case study approach.

We agree that this type of activity would be the most appropriate action at this time to carry out the intent of section 430 of the VA/HUD appropriations bills, as the Academy has already addressed some of the analysis requested in the February 1999 Results Act report. In addition, such a study will create opportunities for dialogue between the agencies, Congress, the Administration and the research community.

We appreciate your efforts on this important issue and look forward to seeing the results.

Sincerely,

F. JAMES SENSENBRENNER, JR.
Chairman

GEORGE E, BROWN, JR.
Ranking Democrat

NAS copy: Rich Bissell, Debbie Stine,
Bill Colglazier, Jim Jensen

United States Senate

WASHINGTON, DC 20510

June 2, 1999

Dr. Bruce Alberts
President
National Academy of Sciences
2101 Constitution Avenue, NW
Washington, DC 20418

Dear Dr. Alberts:

Section 430 of the "VA-HUD-Independent Agencies Appropriations Act for Fiscal Year 1999" requests that the Office of Science and Technology Policy (OSTP), in consultation with the Office of Management and Budget (OMB), enter into an agreement with the National Academy of Sciences to conduct a "comprehensive study to develop methods for evaluating federally-funded research and development programs." That section, a copy of which is enclosed, is based on provisions in the "Federal Research Investment Act," a bill we sponsored in the 105th and 106th Congresses.

We have reviewed the April 6th letter to you from Dr. Neal Lane, the Director of OSTP, regarding the study requested by section 430. In his letter, Dr. Lane asked the Academy to independently assess the strategic and performance plans federal agencies have developed and the responsiveness of their performance reports to the Government Performance and Results Act (GPRA) via a case study approach. This will allow the Academies to suggest specific applications of recommendations from its report "Evaluating Federal Research Programs: Research and the Government Performance and Results Act," which was released in February of this year.

We agree that this type of study would be an excellent way to carry out the intent of paragraph 2 of section 430, especially as paragraphs 1 and 3 of section 430 were largely addressed in the Academies' February report. Thus, we believe the most useful activity at this point would be to carry out the study requested by Dr. Lane, and thus create opportunities for greater dialogue on GRPA among the agencies, Congress, OMB, OSTP, and the research community.

However, we would also ask that you respond to paragraphs 4 and 5 of section 430 as well. In the case of paragraph 4, we request that you undertake an historical analysis of the peer review process to gain a better understanding of the impact that process has had upon whether federal research programs and projects should be funded or continue to be funded. We understand the Academy is currently considering a study of the peer review process; such an analysis would be of great benefit in addressing the issues described in paragraph 4.

For paragraph 5, we believe additional analysis and dialogue is warranted regarding quantitative indicators for program management and administration. While the results of research can be extremely difficult to quantify, particularly in the near term, many aspects of research program management are amenable to quantitative measures. For example, how long does an agency take to make a grant? Or, how high are the administrative costs of a project compared to costs for the research itself?

We appreciate your efforts on this important issue and look forward to seeing the results. We would also appreciate knowing what actions the Academies will take regarding paragraphs 4 and 5 of section 430.

Sincerely,

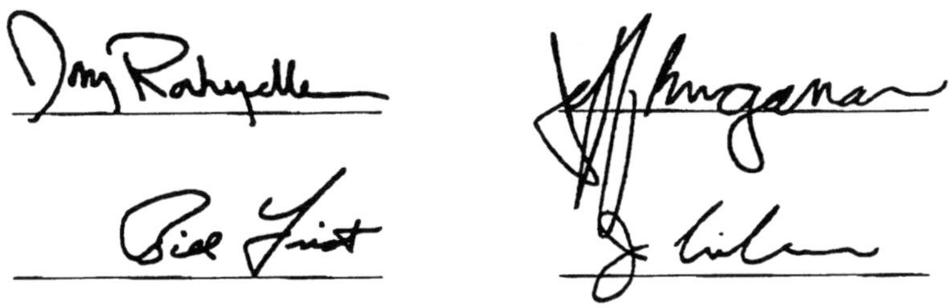

CONFERENCE REPORT ON HR4194, DEPARTMENTS OF VETERANS AFFAIRS AND HOUSING AND URBAN DEVELOPMENT, AND INDEPENDENT AGENCIES APPROPRIATIONS ACT, 1999

SEC. 430. COMPREHENSIVE ACCOUNTABILITY STUDY FOR FEDERALLY-FUNDED RESEARCH.

(a) STUDY.-The Director of the Office of Science and Technology Policy, in consultation with the Director of the Office of Management and Budget, may enter into an agreement with the National Academy of Sciences for the Academy to conduct a comprehensive study to develop methods for evaluating federally-funded research and development programs. This study shall-

(1) recommend processes to determine an acceptable level of success for federally-funded research and development programs by-

(A) describing the research process in the various scientific and engineering disciplines;

(B) describing in the different sciences what measures and what criteria each community uses to evaluate the success or failure of a program, and on what time scales these measures are considered reliable-both for exploratory long-range work and for short-range goals; and

(C) recommending how these measures may be adapted for use by the Federal Government to evaluate federally-funded research and development programs;

(2) assess the extent to which agencies incorporate independent merit-based evaluation into the formulation of the strategic plans of funding agencies and if the quantity or quality of this type of input is unsatisfactory;

(3) recommend mechanisms for identifying federally-funded research and development programs which are unsuccessful or unproductive;

(4) evaluate the extent to which independent, merit-based evaluation of federally-funded research and development programs and projects achieves the goal of eliminating unsuccessful or unproductive programs and projects; and

(5) investigate and report on the validity of using quantitative performance goals for aspects of programs which relate to administrative management of the program and for which such goals would be appropriate, including aspects related to-

(A) administrative burden on contractors and recipients of financial assistance awards;

(B) administrative burdens on external participants in independent, merit-based evaluations;

(C) cost and schedule control for construction projects funded by the program;

(D) the ratio of overhead costs of the program relative to the amounts expended through the program for equipment and direct funding of research; and

(E) the timeliness of program responses to requests for funding, participation, or equipment use.

(b) INDEPENDENT MERIT-BASED EVALUATION DEFINED.-The term "independent merit-based evaluation" means review of the scientific or technical quality of research or development, conducted by experts who are chosen for their knowledge of scientific and technical fields relevant to the evaluation and who-

(1) in the case of the review of a program activity, do not derive long-term support from the program activity; or

(2) in the case of the review of a project proposal, are not seeking funds in competition with the proposal.

APPENDIX C

SUMMARIES OF AGENCY FOCUS GROUP PRESENTATIONS

The following summaries are based on five focus groups held with the major research-supporting agencies during the fall of 2000 and a workshop hosted by the National Academies on December 18-19, 2000. Each focus group was attended by panel members (three of whom were also members of COSEPUP), by agency representatives who were senior research administrators responsible for GPRA compliance, and by representatives of oversight bodies (Congress, OMB, and GAO) responsible for review of GPRA performance plans and reports from research agencies.

A similar agenda was followed during each focus group. The panel began by explaining its goals, agency representatives described their research programs and their mechanisms for GPRA compliance, panel members and oversight representatives commented on agency methodology, and all participants concluded by offering summary comments. The goal of each discussion was to identify aspects of the methodology that could become "best practices" for use by other agencies and areas where the methodology could be improved.

After each focus group, a summary was produced that used the comments and written materials to answer the following questions:

- What methodology is used for evaluating research programs under GPRA?
- What level of unit is the focus of the evaluation?
- Who does the evaluation of the research program under GPRA?

- What criteria are used for the evaluation?
- How do the selection and evaluation of projects relate to the evaluation of the research program?
- How is the result communicated to different audiences (e.g., the S&T community, advisory committees, agency leadership, the administration, Congress)?
- How is the result used in internal and external decision-making?

The workshop was structured differently for a much larger group. The first day's discussion was open to the public and attended by nearly 30 participants. The agenda included a general discussion, an overview, general comments from stakeholders and agencies, breakout sessions, a second general discussion focusing on conclusions and recommendations, panel member comments on the draft report, and a summary session. The second day of the workshop was reserved for panel members, who developed conclusions and recommendations for the report.

SUMMARY OF THE DEPARTMENT OF DEFENSE FOCUS GROUP

1. What methodology is used for evaluating research programs under GPRA?

1.1 Overview.

The Department of Defense (DOD) response to the Government Performance and Results Act (GPRA) is discussed in Appendix I of its Annual Report to the President and the Congress (2000). This appendix summarizes the DOD strategic plan and the ways in which the department links this plan to performance goals and evaluates the performance goals annually.

Virtually all DOD's science and technology (S&T) activities fall under "Performance Goal 2.2 – Transform US Military Forces for the Future." This transforming process is said to be achieved through the development of "new generations of defense technologies." The strategy for achieving these new technologies involves three elements: the basic research plan (BRP), the Joint Warfighting Science and Technology Plan (JWSTP), and the Defense Technology Areas Plan (DTAP).

1.1.1 Basic research. Before World War II, the federal government spent most of its research dollars in federal laboratories. There was considerable opposition to the government's involvement in universities. This was muted by the arguments of Vannevar Bush, who established the conceptual framework for contractual and "unfettered" basic research. Today, DOD invests 56% of its basic research dollars in universities; 29% goes to government laboratories and 11% to industry. Bush argued that such investments in basic research are acts of both faith and patience, but the investments are justified many times over by returns of great value.

DOD's BRP is described as "the cutting edge of the Defense Science and Technology Program." This plan is realized primarily by directly funding research in universities, federal laboratories, and industry and by keeping "a watchful eye on research activities all over the world to prevent technological surprise." The BRP contains an overview of the entire DOD research program, most of which can be described in 12 disciplinary categories.[1] Interdisciplinary research is specifically addressed under three programs. In addition, the plan covers education, training, and instrumentation.

DOD supplies only about 6% of the nation's total federal funding for basic research,[2] but this effort is focused in a number of fields that are critical to the nation's scientific performance. Universities receive 22% of their basic research funding for mathematics from DOD, 42% for computer science, 71% for electrical engineering, 63% for mechanical engineering, and substantial amounts in optics, materials, and oceanography.

[1]Physics, chemistry, mathematics, computer sciences, electronics, materials science, mechanics, terrestrial sciences, ocean sciences, atmospheric and space science, biologic sciences, and cognitive and neural science.

[2]FY2000 federal funding of basic research by funding agency was allocated as follows: National Institutes of Health, 50%; National Science Foundation, 13%; Department of Energy, 12%; National Aeronautics and Space Administration, 13%; DoD, 6%; other, 6%.

1.1.2 Applied research and advanced technology development. The BRP is coupled with two complementary plans that focus on applied research and advanced technology development: the Joint Warfighting Science and Technology Plan (JWSTP) and the Defense Technology Area Plan (DTAP).

The JWSTP takes a joint perspective horizontally across the applied research (6.2) and advanced technology development (6.3) investments to ensure that needed technology and advanced concepts are supported.

The DTAP presents the DOD objectives and the 6.2-6.3 investment strategy for technologies critical to DOD acquisition plans, service warfighter capabilities, and the JWSTP. It also takes a horizontal perspective across the Defense agencies to chart the total DOD investment for given technologies.

1.1.3 DTOs. DOD uses defense technology objectives (DTOs) to provide focus for the development of technologies that address identified military needs across the department. Each DTO identifies a specific technology advancement that will be developed or demonstrated, with expected date of availability, specific benefits resulting from it, and the amount of funding needed. The DTO process is used to comply with GPRA. The output of this process includes budget and management decisions.

1.1.4 TARA. The methodology used for evaluating S&T programs is known as technology area reviews and assessments (TARA). TARA is the department's official response to GPRA, and it is a mechanism to evaluate science and technology programs through expert peer reviews. But in this process, basic research is not isolated from applied research and advanced technology development. All three categories—6.1 (basic research), 6.2 (applied research), and 6.3 (advanced development)—are evaluated as overlapping parts of the technology area being reviewed. For example, biomedical research and chemical-biologic warfare research both have basic-research funding that is particular to them, but they are evaluated in their totality with clear links to what discoveries are expected.

1.1.5 Reliance. The department also uses a process called reliance to guide corporate planning and assessment. Reliance members include the deputy under secretary of defense (science and technology), representatives of all the services, and defense agencies. The objective of reliance is to coordinate the S&T process, stimulate communication among the different services and other groups, and clarify priorities. This is the vehicle for planning and overview that brings the services together. Reliance is designed to encourage collaboration and communication and prevent unnecessary duplication. The group reviews the DTOs themselves, and at the end of the review process all participants sign off on the results of their discussions, so they all have a stake init.

1.2 What level of unit is the focus of the evaluation?

DOD evaluates its S&T activities by reviewing performance at the level of DTOs. There are approximately 400 DTOs, each of which identifies a specific anticipated technology advance, the date of availability, benefits, technologic barriers, and customer. The DTOs are supported by virtually all the S&T defense agencies and services.

1.2.1 The evaluation process. The objectives of the DTAP include creation of technologies that enhance the nation's future warfighting capability. Performance under DTAP can be evaluated by the TARA. TARAs are held every two years for a particular technology area. This year, evaluations are being done in biomedical, battlespace environments, ground/sea vehicles, materials and processes, space platforms, chemical and biological defense, and sensors, electronics and electronics warfare. TARA reviews all three levels of S&T investment—6.1, 6.2, and 6.3.

The TARA reviews are conducted over a period of one week. A review team is asked to evaluate progress toward the individual objectives of DTOs and tries to determine whether that progress should be given a grade of green, yellow, or red.[3] The team is also asked whether a certain area—say, "Detection"—is addressing most of the technologic issues that need to be addressed. Is the research portfolio appropriate for the objective? If part of the program took a serious funding reduction, was the reduction justified? The TARA teams evaluate the programs for quality, for advances in state-of-the-art research areas, and for their scientific vision. Last year, 96% of the department's DTOs were given the grade of green.

1.2.2 Examples of evaluation by DTOs. The following two examples from the 2000 Chemical and Biological Defense Science and Technology TARA illustrate how evaluation works by using the DTOs as the unit of focus. For example, the TARA process gave the "Modeling and Simulation" DTO a yellow grade because of management problems. Because virtually all other DTOs were awarded greens, this was deemed serious enough to trigger a management reorganization. The DTO on "Force Medical Protection": got a red grade because the TARA panel determined that poor technical assumptions and decisions had been made and that researchers were investigating a technology that was not appropriate for the desired objective. As a result, the defense organization performing the work has altered the technical approach to the objectives.

[3]Green means that a program is "progressing satisfactorily toward goals"; yellow means "generally progressing satisfactorily, but some aspects of the program are proceeding more slowly than expected"; red means it is "doubtful that any of the goals will be attained." These DTO ratings are described as semiquantitative metrics that reflect the opinions of independent experts.

Sometimes, such questions are referred for more basic research before major changes are made. In a final example of DTAP DTOs, "Laser Standoff Chemical Detection Technology" received a yellow grade because reviewers decided that the project might, given current performance, have problems after 3 or 4 years. The basis for this judgment was that the project's objective was written in a way that didn't match well with what the researchers were actually doing.

1.2.3 A rationale for "holistic" evaluations.

This process of evaluating performance by DTOs was established before the passage of GPRA, and the management and reporting chains have remained the same. The 6.1, 6.2, and 6.3 aspects of the DTO are all looked at by the same reviewing panel. Panels do not look at the 6.1 element independently, because it is assumed that basic research has important feedback loops with both the applied research and advanced technology development stages.[4]

As an example, DOD is seeking a vaccine for the Ebola virus, and until the basis for such a vaccine is discovered, the research will be funded under the 6.1 category. If a potential vaccine construct is discovered, the vaccine will move to application and development stages, where it will be funded under 6.2 and 6.3 categories. As application and development proceed, further work with 6.1 funds might be needed to achieve a more complete basic understanding and more effective application. Under this same holistic approach, the "Laser Standoff" will be funded under 6.1; if

[4]"Although the DOD model of the transition path from basic research (6.1) to applied research (6.2) to advanced development (6.3) implies a linear model, this is often honored more in the breach than the practice. The 'push' of the linear process is augmented in DOD by a feedback process, whereby changing operational requirements and new results from multidisciplinary research continually keep the Basic Research Program on target." DOD Basic Research Plan, 1999, p. I-5.

the discovery proves out and can be applied and developed, the program will be moved to 6.2-6.3 phases.

1.3 Who does the evaluation of the research program under GPRA?

The evaluation of basic and applied research is carried out by both internal agency panels of experts and by TARA review panels. Each panel consists of 10-12 technical experts from academe, industry, and nonprofit research organizations. Most TARA team members are recognized experts from the National Academies, the Defense Science Board, the scientific advisory boards of the military departments, industry, and academe. Each is chaired by a senior executive appointed by the deputy under secretary for S&T.

These teams are asked to evaluate the programs for quality, for advances in leading the state of the art in research areas, and for their scientific vision. The department requires that two-thirds of each panel be experts from outside DOD. One-third of each panel's members are "refreshed" at the time of each reviewing cycle. Most areas have a 2-year reviewing cycle; chemical-biologic defense is reviewed annually per DOD's implementation of P.L. 103-160.

At a higher level, evaluation is overseen by the Defense Science and Technology Advisory Group (DSTAG), which advises the deputy under secretary for S&T. DSTAG is a key decision-making body consisting of representatives of each service and defense agency. DSTAG provides oversight of an integrated S&T strategic planning process and effectively maintains responsibility for the entire S&T program. It oversees the work of the Basic Research Panel, which consists of eight people and must approve the BRP; the 12 technology panels responsible for preparation of the DTAP; and the 13 panels responsible for preparation of the JWSTP. These plans build on but do not duplicate the service-agency S&T plans.

1.4 What criteria are used for the evaluation?

In the broadest sense, all research activities—like any other DOD activities—must be justified under the mission goals of the agency. If a research project cannot demonstrate its mission relevance, it probably will not be funded.[5]

1.4.1 Evaluating performance. Most specifically, the department evaluates success in achieving the performance goals on two levels. At a lower level of aggregation, individual performance measures and indicators are scored at the end of each fiscal year to determine how performance compared with numeric targets set when the budget was submitted.

At a higher level, annual performance goals are evaluated in two ways. First, results for each of the subordinate measures and indicators are evaluated within the context of overall program performance. Second, a determination is made as to whether a shortfall in expected performance for any metric or set of supporting metrics will put achievement of the associated corporate goal at risk. This subjective determination is trend-based and cumulative. A single year of poor performance might not signal that a corporate goal is at risk, although several years of unsatisfactory performance almost certainly will.

1.4.2 Evaluating basic research. At finer levels—for basic research that is expected to lead to new technologies—the department finds that evaluation through the use of metrics is

[5]DOD's mission, as defined in its strategic plan, begins as follows: "The mission of the Department of Defense is to support and defend the Constitution of the United States; to provide for the common defense of the nation, its citizens, and its allies, and to protect and advance US interests around the world." In practice, this mission includes considerable breadth, especially in regard to its Performance Goal 2.2, which is to "transform US military forces for the future." This goal calls for a continued focus on "new generations of defense technologies," which provides the foundation for its extensive S&T program.

difficult or impossible. There is no reliable way to measure the success of basic research in the near term, because its outcomes are by definition unpredictable. There might be no payoff this year, or next year—until suddenly researchers see a new "data point" that can give rise to a whole new industry.

For this reason, the department chooses to demonstrate the value—and outcomes—of basic research through retrospective achievements. The rationale for this is that the most valuable technologies for defense applications have derived from basic research done years or even decades before the first application. Therefore, the causative process can be more clearly illustrated by looking backward than by conjecturing about future results.

According to the BRP, "a retrospective approach is a reminder that many of the technologies we now take for granted were brought about by investing much earlier in basic research." The following examples all resulted largely from timely DOD investments in basic research:

- Owning the Night (night vision technology).
- Precision Guidance for Air Defense Missiles.
- The Airborne Laser.
- The Kalman Filter (more accurate data for navigation, guidance, and tracking).
- The Global Positioning System.

Retrospective studies are intended to build support for the process, not for individual projects. It is not possible to point to the outcome of an on-going individual project.

1.4.3 Education and training. Other criteria used to evaluate programs include education and training. Clearly, human resources are essential to the future strength of DOD. The department funds more than 9,000 graduate fellowships per year, two-thirds as many as the National Science Foundation (NSF).

However, a difficulty emerges in the way the DOD divides expenditures into the three categories called "Today's Force," "Next Force," and "Force After Next." Most of the department's funds go to readiness ("Today's Force"); the next-highest priority is modernization ("Next Force"); "Force After Next," which contains most S&T and education expenditures, receives a very small percentage of FY2000 appropriations for the department. This difficulty can be seen in the current GPRA format for evaluating S&T. One aspect of the problem is that manpower is considered to be "hard-wired" into the budget process, but there is no evaluation of the educational component itself and thus no incentive structure for good teaching, research training, or mentoring. For example, the substantial cuts in the 6.1 budget from 1993 to 1998 brought reductions in the number of the graduate students who could be supported by research grants at universities, but the GPRA process did not report this result. This is especially troubling for such fields as electrical engineering, computer science, and mathematics, where DOD plays a dominant national role in funding and where basic research is needed to maintain the country's leadership in information technology and other emerging fields.

1.4.4 Relevance to mission. For 6.2 and 6.3 research, R&D activities are clearly aligned with DOD objectives through the DTO categories. For basic (6.1) research, the TARA process does not deal explicitly with how the research is relevant to the DOD mission, but relevance is examined at many points. DOD people attend all TARA reviews, and TARA does review the focus, as well as the quality of the BRP.[6]

[6]In the words of the BRP (p. I-5), "Basic research supported by DOD is directed to maximizing the value that is likely to be created by the investment. Value in this context is represented by the enabling technologies that realize the operational concepts and mission goals of *Joint Vision 2010.*"

In addition, relevance is addressed in the internal management processes. The biennial basic-research cycle starts with project-level reviews at the individual research agencies (Army Research Office, Office of Naval Research, and Air Force Office of Scientific Research). These sessions are followed by program-level reviews of the combined research agencies and by preparation of the BRP. The BRP is evaluated by the director of defense research and engineering, with feedback to the agencies after the annual program review. The services and defense agencies also conduct other periodic program reviews to assess quality, relevance, and scientific progress.

1.4.5 Other criteria. Issues of intellectual property, patents, and spin-offs are also considered to be valuable indicators of the quality and relevance of DOD S&T research. Arranging intellectual property ownership occasionally proves difficult, however, and leads to disputes. These disputes can impede collaboration, most often in university-industry partnerships.

1.5 How does the selection and evaluation of projects relate to the evaluation of the research program?

The selection and evaluation of S&T projects, like all DOD activities, are highly mission-oriented. Projects must have clear objectives, and they must deal with products and product development. The users of products are either in house or in some other agency, so they and their mission are well known as well.

S&T research programs are evaluated in the context of the projects to which they contribute. For example, the DTO of "Detection/Strengths" is analyzed by expert peers. Those conducting the review try to capture their expert opinion, and the opinion is supposed to be based only on the information provided by DOD about the program. In-depth technical reviews are done at the manager level of the project. The Army has its own complex way of evaluating research programs at the service level.

The department has struggled to develop the best way to evaluate research in light of GPRA. When Congress directed the reorganization of DOD in 1987, it put civilian heads in key positions to assess each service and suggest modifications. It was their task to convince military leaders of the value of S&T to the department. The civilian leaders came up with the technique of using planning documents and planning by objective, and an S&T master plan was created in 1990. The introduction of GPRA in 1993 brought a new challenge, and once again the civilian leaders had to make the case for the importance of S&T to the defense mission. Thus evolved the BRP and the two documents that relate it to warfighting objectives.

2. How is the result communicated to different audiences (e.g., S&T community, advisory committees, agency leadership, Administration, Congress)?

The results of TARA reviews are communicated to agency leadership by "TARA outbriefings" for each technology area (6.2 and 6.3) and for basic research (6.1). This provides an efficient way to respond to queries, doubts, or challenges about the value of S&T.

Because GPRA is outcome-oriented and because the TARA mechanism is the department's best process for measuring outcomes, TARA is the best way to communicate the value of what is done by scientists and engineers. More broadly, the department finds TARA so effective as a means of evaluation and communication that it would keep it regardless of GPRA. Within the department, planners struggle through difficult debates and fundamental conflicts about the value of different research programs. The TARA process provides reference points and a means to refer back and forth to areas of planning.

The department also communicates the value of S&T to Congress in several ways, including the use of historical vignettes that demonstrate the utility of basic research. These vignettes are not in the GPRA document at present. In addition, the department communicates with oversight agencies, such as the Office of Management and Budget and the General Accounting Office, about the process of complying with GPRA and the results.

The department also communicates the results of its evaluations to other audiences. For example, nanotechnology research is closely coupled with research supported by NSF, and the National Institute of Standards and Technology. There is strong communication among interagency teams working on national security issues.

Overall, GPRA has the potential to provide a common language to address complex issues and to talk to stakeholders. At present, the various agencies are searching for a common approach that is not yet fully developed.

3. How is the result used in internal and external decision-making?

The TARA review process is used at all levels of decision-making. For example, the TARA 2000 review of chemical-biologic defense S&T program (against such agents as mustard gas, nerve gas, anthrax, plague, and smallpox) revealed that the program was not adequately represented by a DTO portfolio. It also revealed capacity limitations in laboratory infrastructure. In addition, the workforce was observed to be aging. These results will all influence decision-making at the planning and review stages.

TARA panels are able to find redundancies in programs. For example, a panel reviewing a plan for a particular kind of DNA research could point out that this work is already being done by people in industry and recommend that DOD put its dollars elsewhere.

Such strategic decision-making occurs at many levels. The reliance group does a conscious search to ensure that the same work is not being done in two places. For medical projects, the Nationwide Interagency Group helps to prevent duplication by tracking major programs. Learning technology is monitored for duplication by a variety of interagency groups. Basic research in cognitive science is overseen by a combination of NSF and the Office of Science and Technology Policy in the White House.

Other important decision-making that is not addressed by GPRA concerns the choice of basic research fields and the transition of a 6.1 program to a 6.2 and 6.3 program. For example, if a basic research program is shifting its emphasis from high-temperature superconductivity to nanotechnology, researchers know that what they have learned in the former field will benefit their work in the latter. But there is no way to quantify the value of this crossover effect. That kind of flexibility is crucial in basic research as the department seeks to free up money for projects that have the greatest potential to contribute to the defense mission, but it does not appear in the form of metrics. Similarly, quantifying a decision to move a research project from 6.1 to 6.2 is difficult. Such transitions are similarly essential to the mission, and it would be useful to be able to quantify the process. At the same time, the arbitrary application of metrics should be avoided when there is a risk of terminating a potentially useful line of inquiry.

DOD Focus Group Participant List
October 4, 2000

Panel Members:

Enriqueta Bond
President
The Burroughs Wellcome Fund
Research Triangle Park, North Carolina

Alan Schriesheim
Director Emeritus
Argonne National Laboratory
Argonne, Illinois

John Halver
Professor Emeritus in Nutrition
School of Fisheries at University of Washington
Seattle, Washington

Morris Tanenbaum
Retired Vice Chairman and
Chief Financial Officer, AT&T
Short Hills, New Jersey

Robert M. White
University Professor and Director
Data Storage Systems Center
Carnegie Mellon University
Pittsburgh, Pennsylvania

Participants:

James Baker
Internal Medicine
University of Michigan
Ann Arbor, Michigan

Greg Henry
Program Examiner, National Security Division
Office of Management and Budget
Washington, D.C.

Genevieve Knezo
Congressional Research Service
Library of Congress
Washington, D.C.

Steven Kornguth
University of Texas at Austin
Institute for Advanced Technology
Austin, Texas

David Lockwood
Congressional Research Service
Library of Congress
Washington, D.C.

Elizabeth Mead
Senior Analyst
US General Accounting Office
Washington, D.C.

Jack Moteff
Congressional Research Service
Library of Congress
Washington, D.C.

Robin Nazzarro
Assistant Director
US General Accounting Office
Washington, D.C.

Michael Sailor
Department of Chemistry
University of California, San Diego
La Jolla, California

David Trinkle
Program Examiner, Science and Space Programs Branch
Office of Management and Budget
Washington, D.C.

Agency Representatives:

Robert Foster
Director, Biosystems
US Department of Defense
Roslyn, Virginia

Joanne Spriggs
Deputy Under Secretary of Defense (S&T)
US Department of Defense
Washington, D.C.

Robert Trew
Director of Research
US Department of Defense
Washington, D.C.

Robert Tuohy
Director, S&T Plans and Programs
US Department of Defense
Roslyn, Virginia

Leo Young
Office of Director for Research
US Department of Defense
Arlington, Virginia

SUMMARY OF THE NATIONAL INSTITUTES OF HEALTH FOCUS GROUP

1. What methodology is used for evaluating research programs under GPRA?

1.1 Overview.

The National Institutes of Health (NIH) is an agency within the Department of Health and Human Services (DHHS). NIH's mission is to uncover new knowledge and to develop new or improved methods for the prevention, detection, diagnosis, and treatment of disease and disability. Preparation of NIH's annual GPRA performance plans and performance reports is the responsibility of the Office of Science Policy within the Office of the Director, NIH. GPRA documents are formally submitted by NIH, through DHHS, in conjunction with the normal cycle of budget document submission throughout the year. In compliance with the requirements of GPRA, NIH has prepared and submitted Annual Performance Plans for Fiscal Years 1999, 2000, and 2001. The FY 2002 Performance Plan is now being developed.

Like other federal agencies that support scientific research, NIH faced challenges in evaluating the outcomes of its research programs in accordance with the requirements of GPRA. Compliance with GPRA required the NIH to implement an assessment and reporting process that complemented the ongoing mechanisms for review of research progress.

1.1.1 NIH GPRA Performance Plan. For purposes of GPRA planning and assessment, the NIH has aggregated and categorized the mission-related activities of all its Institutes, Centers, and Offices into three core program areas: Research, Research Training and Career Development, and Research Facilities. For each of these three core program areas, the NIH has identified expected outcomes, major functional areas, specific performance goals, and annual targets within its GPRA performance plan. The performance goals in NIH's annual performance plans address both the long-term, intended results or outcomes of NIH core program activities and the management and administrative processes that facilitate the core program activities and lead to the achievement of outcomes. For example, within the Research Program, outcome goals include increased understanding of biological processes and behaviors, as well as the development of new or improved methods for the prevention, diagnosis, and treatment of disease and disability. NIH's Annual Performance Plans include performance goals that can be assessed through the use of objective/quantitative measures as well as performance goals that require descriptive performance criteria.

1.1.2 Quantitative measures. Most of the 50-odd performance goals described in NIH's annual performance plans can be assessed through the use of objective and quantitative measures, such as numerical targets, data tracking and collection systems, completion of studies or actions, and program-evaluation studies.

For example, two of the seven primary research goals can be evaluated quantitatively. One is to develop critical genome resources, including the DNA sequences of the human genome and the genomes of important model organisms and disease-causing micro-organisms. An example of a quantitative goal for FY2001 was to complete the sequencing of one-third of the human genome to an accuracy of at least 99.99%. A second quantitative goal for FY2001 was to work toward the president's goal of developing an AIDS vaccine by 2007; progress is described in terms of the design and development of new or improved vaccine strategies and delivery or production technologies. For both goals, NIH was able to identify specific milestones or other measurable targets.

1.1.3 Qualitative criteria. The annual performance goals related to its dominant research program, however, are more qualitative. NIH has concluded that strictly numeric goals and measures are neither feasible nor sufficient to capture the breadth and impact of research it performs and supports. In such cases, GPRA provides an avenue for an agency to define performance goals that rely on criteria that are more descriptive and to use an alternative form of assessment.

A small subset of the annual performance goals, related to the NIH Research Program, are more qualitative in nature, and the NIH has used the alternative form, as allowed by the GPRA, for these five goals:

- Add to the body of knowledge about normal and abnormal biological functions.
- Develop new or improved instruments and technologies for use in research and medicine.
- Develop new or improved approaches for preventing or delaying the onset or progression of disease and disability.
- Develop new or improved methods for diagnosing disease and disability.
- Develop new or improved approaches for treating disease and disability.

For the five qualitative goals mentioned above, an independent assessment process has been developed and is described in more detail below.

1.2 What level of unit is the focus of the evaluation?

For purposes of strategic planning under GPRA, an agency is defined as a cabinet-level department or independent agency. In the case of NIH, this means that the parent agency, DHHS, is the agency that must develop a long-term strategic plan. NIH's core programs support the strategic plan of DHHS, and NIH provides considerable input into its development. Each of the NIH operating units has its own plan and reports in a formal way through the department. These units may be formed around a disease, a phase of the human life cycle, a biologic system, or a profession (such as nursing). The overall DHHS performance plan is the total of the plans of its 13 subagencies, which in turn stem from the DHHS strategic plan.

The primary dilemma of those in charge of the response to GPRA has been the size and complexity of NIH itself. For example, they found no meaningful way to evaluate the research results of individual institutes and centers, because the work of each unit overlaps with and complements the contributions of others. And each institute and center has its own strategic plans, devised independently.

1.2.1 Aggregation. For a broader view of GPRA assessment, NIH has chosen the option of aggregating the activities of its institutes, centers, and offices (as permitted by GPRA) in the three core program areas described above.

1.3 Who does the evaluation of the research program under GPRA?

During NIH's planning for GPRA, there was considerable discussion about who should do the assessment. NIH officials decided that the evaluation should be conducted by an Assessment Working Group of the Advisory Committee of the Director (ACD), NIH, following the highly effective model of peer review that is used by NIH for merit review of grant applications. This Working Group drew its membership from members of the ACD, the Director's Council of Public Representatives, and members of NIH Institute and Center national advisory councils that provide advice on a broad range of topics. This combination of individuals provided broad representation of the scientific and medical communities, health care providers, patients, and other representatives of the public. Moreover, it provided the expertise and perspectives necessary for evaluating the scientific quality and societal relevance of the outcomes of the NIH Research Program. The final working group had 26 members: six ACD members, 16 COPR members, and four ad hoc scientists selected by the NIH director for their scientific expertise in areas not already represented.

1.4 What criteria are used for the evaluation?

1.4.1 Peer review. One reporting challenge for NIH is that most of its funding for research does not stay within the system. Some 82% of the budget goes outside NIH in the form of extramural awards, compared with about 11% that pays for intramural research at the Bethesda, MD, campus and other centers. Each year, NIH receives some 40,000 research proposals from scientists and research centers throughout the country. Much of the extramural research is performed by principal investigators at universities who employ and train graduate students and postdoctoral scientists on their grants. These projects, which might have multiple funders, are not under the direct control of NIH. They are, however, governed by an effective and long-standing peer-review process with stringent requirements and evaluation procedures. NIH found no need to attempt to duplicate or replace this system, which is the traditional means of research approval and assessment used in scientific programs.

Therefore, to develop its approach to GPRA, NIH developed an independent assessment process for evaluating program outcomes and compared them with the performance goals for the research program. In the broadest terms, the assessment involved "gauging the extent to which NIH's stewardship of the medical research enterprise leads to important discoveries, knowledge, and techniques that are applied to the development of new diagnostics, treatments, and preventive measures to improve health and health-related quality of life" (from the NIH GPRA Research Program Outcomes for FY1999).

1.4.2 Assessment materials. The working group was provided with narrative "assessment materials" that consisted of the following evidence of research-program outcomes:

- Science advances. One-page articles prepared by NIH that describe a specific scientific discovery published within the last year and supported by NIH funding. Each advance was related to its impact on science, health, or the economy.[1]
- Science capsules. One-paragraph snapshots of the breadth and scope of individual NIH research program outcomes. Their brevity allows for a greater number of vignettes, each offering a thumbnail description of an advance and its significance, so that the overall picture created by the capsules is more nearly representative of the research effort as a whole.[2]
- Stories of discovery. One- to two-page narratives that trace a major development over several decades of research, demonstrating the long-term, incremental nature of basic research and its often-surprising utility in seemingly unrelated areas of medicine. These narratives address the difficulty of attempting to

[1]Typical topics of the science advances were "Enzyme Can Repair Alzheimer's Tangles," "Pathways in the Brain That Control Food Intake," and "Proteins as Genetic Material in Human Disease."

[2]Typical topics of the science capsules were "The Brain's Capacity to Change," "Understanding Cataract Formation," and "Homocysteine: Another Kind of Heart Risk."

describe important advances in terms of a single finding or annual accomplishments.[3]

- Research awards and honors. Brief descriptions of national and international scientific awards or honors received by NIH scientists and grantees. The write-ups demonstrate how the external scientific community values the work of NIH grantees.

Narrative descriptions of research accomplishments were accompanied by citations of publications related to the accomplishments.

To assemble the narrative materials about outcomes, each NIH institute and center was asked to provide 10-20 science advances, 10-20 science capsules, and one or two stories of discovery. The resulting assessment materials were considered to provide an extensive illustration of NIH's FY1999 research outcomes that address the five qualitative research-program performance goals.

1.4.3 Evaluating outcomes. A total of almost 600 advances, capsules, and stories of discovery were given to the working group 3 weeks before its 1-day assessment meeting. For the meeting, each member was asked to review a subset of the materials: those for goal A ("add to the body of knowledge..."), those for one additional goal (instruments and technologies, prevention, diagnosis, or treatment), and the research awards. Each was asked to identify, if possible, some five noteworthy scientific discoveries from each assigned goal and to identify any findings considered "marginal."

At the 1-day meeting, the working group discussed in plenary session the research outcomes for goal A and discussed and assessed goals B through E (instruments and technologies, prevention, diagnosis, and treatment) in breakout groups.

[3]Typical topics of the stories of discovery were "Drug Exposed Children: What the Science Shows," "Challenging Obesity," and "Helping Couples Conceive."

After the meeting, the working group was asked to evaluate the outcomes for each goal. To assess goal A, the working group was asked to use the following criteria:

- The NIH biomedical research enterprise *has successfully met this goal* when its research yields new findings related to biologic functions and behavior, and the new findings are publicized or disseminated.
- The NIH biomedical research enterprise *has substantially exceeded this goal* when, in addition to fulfilling the above criteria, any of the following apply:

  —Discoveries result in significant new understanding.

  —Research yields answers to long-standing, important questions.

  —Genome information about humans, model organisms, or disease-causing agents is translated into new understanding of the role of genes or the environment.

  —Discoveries have potential for translation into new or improved technologies, diagnostics, treatments, or preventive strategies.
- It was also explicitly pointed out to the working group that a third level of performance—*the goal was not met*—was also possible and could be considered.

1.4.4 Specifying results. The compilation of written materials for goal A was by far the largest of that for any goal, totaling 265 items. For this goal, the working group concluded that NIH had "substantially exceeded" the goal of "adding to the body of knowledge." Specifically, the working group concluded that the outcomes demonstrated that NIH had "sustained the excellence and responsiveness of the research system while demonstrating willingness to take research risks necessary to advancing biomedical knowledge and, ultimately, human health."

In all, the group judged that for FY1999, NIH had "substantially exceeded" four goals and "met" one goal. The goal that "lagged" somewhat was goal C, "Develop new or improved approaches for preventing or delaying the onset or progression of disease and disability."

1.4.5 COSEPUP criteria. At the workshop, there was little discussion of one of the three COSEPUP criteria for evaluating research—that of leadership. The other two criteria—quality and relevance to mission—were either discussed at length (quality) or embedded in the peer-review process (relevance to mission). Leadership concerns the relative level of research being performed in a given program relative to the highest world standards of excellence. COSEPUP has suggested, and tested, the use of "international benchmarking" to measure leadership, a technique discussed in its full report.[4]

1.5 How does the selection and evaluation of projects relate to the evaluation of the research program?

Selection criteria were developed by NIH on the basis of the decision to aggregate its individual research projects and to evaluate them as part of broad biomedical goals. The objective of these criteria was to capture the results of clinical, as well as basic, research. NIH staff held many roundtable discussions, conferences with stakeholders, and cross-agency planning sessions to gather input from all groups. They used the National Association of Public Administration (NAPA) as a forum.

1.5.1 Research as the primary mission. NIH is in the midst of a planning-while-doing process to find the best way to evaluate a research-dominated mission. Most mission-based agencies—such as the Department of Defense, the Department of Energy, and the National Aeronautics and Space Administration—

[4] *Evaluating Federal Research Programs.*

spend only a small fraction of their budgets on research, and their missions are described in terms that are not restricted to research (such as maintaining the national defense or exploring the solar system). NIH, in contrast, like the National Science Foundation, has research as its primary mission and allocates its budget accordingly. Therefore, the evaluation of its "performance and results" is primarily a matter of evaluating the research effort itself.

1.5.2 Reviewing basic research. The reasons for this approach are derived from the unique challenges for agencies whose missions include basic and clinical research. As proposed in NIH's report, *Assessment of Research Program Outcomes*, scientists and the practice of science "exist because of what we do not know. The aim of science is to move the unknown into the realm of the known and then, with a greater store of knowledge, begin again, as if advancing a frontier. This basic truth about science makes it different from other enterprises."

Because it is impossible to know with certainty which field or project of research will produce the next important discovery, the assessment report continues, the community of science "has to be open to all ideas." Many of these ideas will lead to useful outcomes; many others will not. Although much NIH funding supports research projects that are of obvious relevance to specific diseases and public health, it also places a high priority on fundamental, untargeted research. History has shown many times that a basic-research finding might be a critical turning point in a long chain of discoveries leading to improved health and richer science. However, although these basic research programs can be evaluated usefully on a regular basis, the ultimate outcomes of fundamental research are seldom predictable or quantifiable in advance.

1.5.3 Dealing with unpredictability. According to the NIH assessment report, unpredictability has three important implications:

- Science is by nature structured and self-correcting, so that either a predicted or an unforeseen discovery has the advantage of adding to basic scientific knowledge and giving new direction to further inquiries.
- Science and its administrators must constantly reevaluate and often change their priorities in light of new discoveries.
- Tracking the many aspects of fundamental science is a daunting challenge that must capture quantitative, qualitative, and institutional dimensions... . It is normal and necessary for basic research investigators to modify their goals, change course, and test competing hypotheses as they move closer to the fundamental understandings that justify public investment in their work. Therefore it is necessary to evaluate the performance of basic research programs by using measures not of practical outcomes but of performance, such as the generation of new knowledge, the quality of the research performed, and the attainment of leadership in the field.

In addition, the annual reporting requirements of GPRA present a problem. The outcomes of fundamental science commonly unfold over a period of years. During a given period, it might or might not be possible to quantify progress or predict the direction of the outcome.

2. How is the result communicated to different audiences (e.g., S&T community, advisory committees, agency leadership, Administration, Congress)?

An agency's response to GPRA can enhance communication both internally and externally. Internally, the exercise at NIH has focused attention on how the institutes manage their activities. It has required a common dialogue in the parent agency, DHHS. It forces both the agency and NIH to set goals and be accountable. NIH's partnership with other agencies, such as NSF and its association with NAPA has brought all participants a somewhat better understanding of Congress's interest in GPRA, and it will be used.

2.1 *Toward common nomenclature.*

One impediment to internal communication in the past has been the use of different standards and nomenclature. GPRA encourages common standards and nomenclature during all phases of the process. Improved communication between biomedical disciplines is more important today than in the past. A generation ago, there was a clear line between research and care. Today, the pace of research is greater, and the line between research and care is blurred. For example, virtually all pediatric cancer care now uses research protocols, as does a growing proportion of adult cancer care.

2.2 *Communicating with oversight bodies.*

A primary audience is Congress. It is still unclear how GPRA information is being used by the authorizing committees or, especially, the appropriations committees, but the House Science Committee has been an active participant in planning and overseeing implementation of GPRA. Another important audience is GAO, which oversees many aspects of government performance. Finally, OMB is both a participant and an audience in the GPRA process for NIH. OMB receives the budget requests and performance plans, engages with the agency, and asks questions about how well the plan reflects the agency's priorities. OMB has learned, along with NIH, how research presents different challenges for evaluation. OMB reports a generally favorable opinion of NIH's aggregation plan.

3. How is the result used in internal and external decision-making?

The GPRA process has greatly facilitated internal decision-making by bringing groups together and establishing linkages throughout DHHS. When the working groups were established, new contacts and interoffice relationships were built. People learned how different institutes, centers, and agencies had different approaches to planning and reporting. Groups were able to look at different plans and understand them. They also learned how there could be a combined plan and report.

DHHS is attempting to use the results to improve the linkage between performance plans and budget. The linkages are not made dollar for dollar, especially in research, but the information gathered for GPRA is useful to help make decisions earlier in the budgeting process. Performance plans are also used by the DHHS budget review board, which has made a commitment to using GPRA. Ultimately, planners hope to use it more explicitly for budgeting, and even for internal management.

3.1 Linkage with the budget.

The use of GPRA activities to feed back into the budget process is complex: how to get from the appropriations process (on an institutional basis) to how each institute and center spends money. Many players are involved in budgeting, including each institute's director, the NIH director, the White House, OMB, Congress, and the President. The appropriations committees have the final decision on budget amounts. Once the money reaches NIH, it must decide how to allocate it. On a grassroots level, individual investigators (both extramural and intramural) guide the disbursement at the micro-level by proposing promising ideas for research. NIH also advertises areas for which it solicits more research (for example, in diabetes, it might call for more work on islet-cell transplantation). The Institutes also try to balance their disbursements between small grants and large projects, between basic and clinical research, and between research and instrumentation or other infrastructure.

The experience with GPRA is too brief to allow NIH to place a value on the process. There has been only one assessment report, and a second is due in spring 2001. GPRA's effect is also hard to discern amid other forms of input. At least, it has proved to offer another avenue of feedback and evaluation.

3.2 Decisions about goals not met.

Failure to meet selected goals would not necessarily trigger a shift in resource allocations. For example, if a target was not met because it was too ambitious, the problem could exist in the target-setting, rather than in inadequate funding or poor execution. For example, NIH set a target for facility construction last year that was not met. The target had been set 2 years before. The reason it was not met was that it turned out to be more cost-effective to add a floor to the building during construction, thereby extending the completion date beyond the initial target. Such situations are not clearly dealt with in some GPRA reporting mechanisms.

3.3 Changing the process.

NIH has made several changes in the assessment process as a result of the previous assessment:

- It has added a co-chair of the working group so that each year's chair will have had the experience of a previous cycle.
- The working group has been expanded (to 34 people) to add expertise and ensure sufficient coverage when some cannot attend.
- Specific review assignments were made to facilitate the assessment process.
- Discussion of each individual's and the group's collective assessment by goal was conducted during the meeting, rather than following the meeting by ballot.

NIH Focus Group Participant List
October 25, 2000

Panel Members:

Enriqueta Bond
President
The Burroughs Wellcome Fund
Research Triangle Park, North Carolina

Alan Schriesheim
Director Emeritus
Argonne National Laboratory
Argonne, Illinois

Brigid L.M. Hogan
Investigator, Howard Hughes Medical Institute and Professor,
Department of Cell Biology
Vanderbilt University School of Medicine
Nashville, Tennessee

Max D. Summers
Professor of Entomology
Texas A&M University
College Station, Texas

Bailus Walker, Jr.
Professor, Environmental & Occupational Medicine
Howard University
Washington, D.C.

Participants:

Theodore Castele, M.D., FACR
Member, NIH GPRA Assessment Working Group
Fairview Park, Ohio

Melanie C. Dreher
Dean and Professor
The University of Iowa College of Nursing
Iowa City, Iowa

Marc Garufi
Program Examiner, Health Programs and Services Branch
Office of Management and Budget
Washington, D.C.

Joanna Hastings
Program Analyst, Office of Budget / ASMB
Department of Health and Human Services
Washington, D.C.

Genevieve Knezo
Congressional Research Service
Library of Congress
Washington, D.C.

Robin Nazzaro
Assistant Director
US General Accounting Office
Washington, D.C.

Robert Roehr
Member, NIH GPRA Assessment Working Group
Washington, D.C.

Arthur Ullian
Member, NIH GPRA Advisory Committee
National Council on Spinal Cord Injury
Boston, Massachusetts

Agency Representatives:

Robin I. Kawazoe
Director, Office of Science Policy
& Planning
National Institutes of Health
Bethesda, Maryland

Lana R. Skirboll
Associate Director for Science Policy
National Institutes of Health
Bethesda, Maryland

John Uzzell
Director, Division of Evaluation
Office of Science Policy
National Institutes of Health
Rockville, Maryland

SUMMARY OF THE NATIONAL AERONAUTICS AND SPACE ADMINISTRATION FOCUS GROUP

1. What methodology is used for evaluating research programs under GPRA?

1.1 Overview.

Like other federal agencies that support significant programs of science and engineering research, the National Aeronautics and Space Administration (NASA) has encountered several difficulties in evaluating research programs to the specifications of GPRA. Compliance with GPRA did not require new mechanisms for assessing performance; these were already in place. But it did require a new overlay of reporting requirements on an annual basis.

For its first performance report, in FY1999, the agency assessed its performance in terms of goals (e.g., "solve mysteries of the universe"), multiple objectives (e.g., "successfully launch seven spacecraft, within 10 percent of budget, on average"), and targets achieved (e.g., "several spacecraft have been successfully developed and launched with a 3.8 percent average overrun").

For its FY2001 performance plan, NASA has instituted several major changes. The targets have been changed in an effort to relate the specific annual measures of output (now called "indicators") to the eventual outcomes that usually result from a number of years of activity. By using the new targets, NASA hopes to have better success in relating multiyear research programs to yearly budget and performance reports. For example, under the strategic plan objective "solve mysteries of the universe" are three "targets" (e.g., "successfully develop and launch no less than three of four planned missions within 10% of budget and schedule"). Under this target are a series of specific performance indicators, such as the successful launch of the microwave anisotropy probe.

1.1.1 Internal and external reviews. To take a broader view of NASA's evaluation techniques, the agency uses extensive internal and external reviews to assess its research efforts against its performance plans. Internal reviews include standard monthly and quarterly project- and program-level assessments at NASA centers, contractor sites, and NASA headquarters. There are reviews of science, engineering, and technology plans and performance, in addition to reviews of functional management activities linked to research, such as procurement, finance, facilities, personnel, and information resources. Management councils conduct oversight reviews of schedules, cost, and technical performance against established plans and bring together headquarters and field directors twice a year for assessment reviews of enterprise and cross-cutting process targets.

When GPRA was introduced, NASA management decided that existing review processes were sufficient to provide internal reporting and reviewing of project and program performance data. The recent streamlining of agency processes provided confidence that new data-collection and oversight processes need not be created for compliance with GPRA.

For external review, NASA relies on National Science Foundation-style peer review of its activities by outside scientists and engineers, primarily from universities and the private sector. Panels of scientific experts are asked to ensure that science-research proposals are selected strictly on their merits. "Intramural" (at NASA facilities) projects in the research programs are selected in the same competitive processes as extramural (e.g., at universities) projects. Competitive merit review is applied to over 80% of resources awarded.

Additional reviews are conducted by such organizations as the NASA Advisory Council (including representatives of universities, industries, and consulting firms), the National Research Council, and the General Accounting Office.

For the purposes of complying with GPRA, NASA relies on its own advisory committees for its primary input. These committees are already familiar with NASA's strategic plan, individual enterprise strategic plans, and budget.

1.1.2 The need to tailor GPRA to particular areas. At the workshop, NASA devoted about half its time to presenting the research programs in the three strategic enterprises with science missions, and the GPRA responses of those specific research programs: space science, earth science, and biologic and physical science. Each has different requirements, and each should have its own methods for complying with GPRA. That is especially true in evaluating internal programs and setting internal priorities. If reviews are effective at allowing the reallocation of dollars, they must reach below the program level to comprehend individual projects. Agencies must stimulate internal discussions among divisions so that peer-review systems can be carefully tailored to match the needs of specific areas.

1.1.3 The need to evaluate technologic activities. An essential part of the "science cycle" for NASA is a wealth of technologic activities that include theoretical studies, new-instrument development, and exploratory or supporting ground-based and sub-orbital research that are intended to help accomplish scientific research objectives. These technology programs, as integral parts of NASA's overall research effort, must be evaluated for GPRA with the same transparency and rigor as its "science" products. In addition, reviewers should assess the effectiveness with which these technologic activities are integrated with "scientific" complements.

1.1.4 Additional challenges. Evaluating research programs under GPRA presents other significant challenges to the space agency. One reason is that it must deal with 3 years of planning and evaluating simultaneously. For example, NASA is currently developing the performance plan for the budget planning year (02), tracking the current plan for the current budget year (00), and preparing the performance report for the completed fiscal year (01).

The development of metrics is also complicated by the issue of lead time. In NASA's earth-science programs, for example, 16 months pass between the time when targets are submitted and the time of implementation; 28 months pass between target submission and final results for the performance report. During those periods, the ideas and basis of the research program often change substantially, forcing alterations of individual metrics and perhaps even of the larger goals of the program.

Finally, perhaps the most difficult challenge is to develop an appropriate response to GPRA's focus on outcome metrics. Historically, NASA is accustomed to tracking the technical status and progress of its complex programs, and accurate tracking is integral to the success of its missions. For flight programs, such as the International Space Station, NASA engineers use many thousands of technical metrics to track performance, schedules, and cost.

For long-term research programs, however, such technical metrics might not adequately convey the quality or relevance of the work itself. For example, in the space-science objective to "solve the mysteries of the universe," the assessment process requires a multifaceted judgment that takes into account the nature of the challenge to "solve the mysteries," the level of resources available, and the actual scientific achievements of the year. That judgment cannot be achieved solely by comparing the number of planned events for the year with the number of events that were achieved.

This issue will be discussed at greater length in Section 1.5 below.

1.1.5 Overall performance. For the purpose of assessing NASA's performance at the enterprise and cross-cutting process levels, reviewers must integrate quantitative output measures and balance them with safety, quality, performance, and appropriate risk. The advisory committees will be asked to assign a rating of red, yellow, or green to measure the progress made against each of the objectives and provide a narrative explanation. These objectives are identified in the strategic plan and repeated in the display of strategic goals and strategic objectives.

1.2 What level of unit is the focus of the evaluation?

NASA divides its activities into five "enterprises": the Space Science Enterprise, the Earth Science Enterprise, the Human Exploration and Development of Space Enterprise, the Biological and Physical Research Enterprise,[1] and the Aero-Space Technology Enterprise. Each enterprise then evaluates is mission by several strategic plan goals. Each goal, in turn, has several strategic plan objectives; each objective has one or more targets for the fiscal year, and each target is measured by one or more indicators, as described above.

For the GPRA assessment of the Space Science Enterprise, for example, there are three components:

- Mission development. This component has about 20 specific targets, from successful launches to specific missions. Each is reviewed for progress in design or for success in bringing technologic development to a certain level.
- Mission operations and data analysis. Independent outside reviewers are asked to evaluate the NASA program with regard specifically to strategic plan goals and generally to how the agency contributes to space science as a whole.

[1]This enterprise was reorganized from the Office of Life and Microgravity Science and Applications (OLMSA) in September 2000. OLMSA was part of the Human Exploration and Development of Space Enterprise.

- Research and data analysis. This component uses independent outside reviewers on a triennial cycle, which is more appropriate than an annual cycle for research. NASA's space science research programs receive about 2000 proposals per year for research grants; of those, it selects 600-700. The proposals are screened with traditional peer review. In addition, the agency has begun an additional layer of expert review called senior review, as recommended by COSEPUP in its GPRA report of February 1999.[2] For this review, instead of looking at 2000 awards in 40 disciplines, NASA has grouped all projects in nine science clusters. Reviewers look at highlights of the clusters and examine recent research accomplishments that meet strategic-plan goals. They also review work in progress that is designed to meet long-term goals. For example, future space missions will require new forms of imaging that must be supplied by basic research in optics; reviewers will monitor NASA's progress in optics research toward this goal.

The Space Science Enterprise initiated a planning process for this mechanism 18 months ago. The first triennial senior review will be held in the middle of 2001 to fit in with the strategic-planning process.

1.3 Who does the evaluation of the research program under GPRA?

Many standing and ad hoc groups participate in the evaluation process. At the enterprise level, the target "owners" are asked for the most appropriate indicators to use as metrics. These metrics are reviewed by the independent NASA Advisory Council. The GPRA coordinators take this input, integrate it with the rest of the performance plan, and send it to the Office of Management and Budget (OMB).

---

[2]*Evaluating Federal Research Programs.*

Oversight of the GPRA process for the NASA science research programs is the responsibility of the NASA Science Council and interagency working groups. The Science Council is an internal group composed of the chief scientist, chief technologist, CFO, and other members of the headquarters leadership. (Because of the relationship between the budget and the performance plans, the NASA CFO has primary responsibility for the conduct of the performance plan and reporting process.)

Several external groups also help to guide the process. The Space Studies Board of the National Research Council performs guidance and evaluation. Other participants include the Institute of Medicine Board on Health Science Policies and the Aeronautics and Space Engineering Board and National Materials Advisory Board of the National Research Council.

1.3.1 An example of the evaluation process.

For FY2002 performance-target development, NASA headquarters transmits guidelines on targets and goals based on GPRA, OMB Circular A-11, and the Congressional Scorecard. The lead NASA centers develop performance targets with additional guidance from advisory committees. Headquarters reviews the targets and develops specific plans. These plans and targets are included in program commitment agreements between the administrator and associate administrator. They are reviewed internally and then by OMB. During the fiscal year, progress toward the targets is reviewed at least quarterly by the NASA Program Management Council in light of both budget development and strategic plans. Final review is conducted by the advisory committee, and the result is a grade of red, yellow, or green.

Peer-review research at the project level does not appear on GPRA plans or reports, because of the great number of projects involved. Nonetheless, it is the fundamental mechanism for maintaining the quality of NASA's research programs. Disciplinary working groups are responsible for overseeing the peer-review process for each discipline and for developing milestones.

For example, in the Earth Sciences Enterprise, essentially all scientific research is peer-reviewed through the use of mail or panel reviews. In FY1990, nearly 5000 mail reviews were received, and peer review panels in FY1999 and FY2000 involved nearly 300 people. Also, NASA Earth Science Enterprise programs have extensive interaction with the community, including National Research Council panels, US Global Change Research Program, and international science organizations (such as World Climate Research Program and EGBP).

1.3.2 Who is "external"? Some aspects of peer review were debated at the workshop, notably the need to use reviewers who are able to evaluate a project objectively. As a rule of thumb, NASA prefers a panel of whom one-third or more are not currently funded by NASA.

1.4 What criteria are used for the evaluation?

As mentioned above, GPRA requires a heavier focus on *outcome* metrics than on NASA's common input and output metrics. For example, OMB Circular A-11 states that performance metrics must be measurable, quantifiable, reflective of annual achievement, and auditable. Like other agencies that support research, NASA has difficulty in finding such performance metrics for research programs and in relating multiyear projects to the annual budget process.

Workshop participants discussed this issue in detail. To distinguish the two terms, an example of an output might be a workshop or a launch—a deliverable that might or might not have value beyond the simple performance of the task. An outcome, in contrast, would be evidence that a workshop increased knowledge or new science enabled by data obtained from a NASA payload in orbit—concrete benefits or results.

1.4.1 Measuring outcomes of research. Because the outcomes of most research programs are not clear for several years, especially those requiring launching, the effort to report outcomes can lead to the use of numbers that mean little with respect to the new knowledge hoped for. Conversely, a program might report successful outputs (e.g., preparation of experiments for launch) that are nullified if the launch fails or is postponed. In other words, it is possible to meet the indicators and miss the target—or to miss the indicators and still learn a great deal about the target objective.

1.4.2 A plan for expert review. For those reasons, NASA is planning to change its reporting process for FY2002. The agency is now evaluating the changes, discussing them internally, and gauging how they will apply to the GPRA process. The struggle is to quantify "intangible" results, such as knowledge. Most government programs have a product that is easy to describe, including many NASA missions. But when knowledge is the objective, its form is unknown, and its discovery is often serendipitous. That kind of objective defies the use of conventional metrics.

Hence, the new process makes use of expert review of the research-program components to attempt a more meaningful approach. NASA will continue to report annual GPRA-type metrics for enabling activities, such as satellite development, as well as annual science accomplishments in the GPRA performance report. It would review one-third of the research program annually, providing regular scrutiny. It would need to ensure a review of the degree of integration within research and the connection of the research to applications and technology. Many NASA centers already do this, but it has not been enterprise-wide, and NASA will have to get its budget office to approve it. Originators of this approach believe that the scientific community will show far more enthusiasm for evaluating research programs with expert review than for evaluation according to annual measures and results. The experience of centers that have used expert review is highly favorable.

This would relieve several major concerns about the past method. One is concern that when the importance and relevance of a program are defined in terms of metrics, a program considered unmeasurable or difficult to measure could lose priority in the budget process relative to programs that are easier to quantify. Similarly, unmeasurable or difficult-to-measure programs give the perception that their progress and ability to produce useful results are not being tracked regularly. The use of expert review to track program performance could be accurately reflected in the performance report.

1.4.3 A plea for change. COSEPUP, in its 1999 report on GPRA, suggested that "there are meaningful measures of quality, relevance, and leadership that are good predictors of eventual usefulness, that these measure can be reported regularly, and they represent a sound way to ensure that the country is getting a good return on its basic research investment."

1.5 How does the selection and evaluation of projects relate to the evaluation of the research program?

One way to describe the relationship between the selection and evaluation of projects and the evaluation of the research program is in terms of relevance. That is, a research project should be selected only if it is relevant to the long-term goals of the research program. That is the difference between doing research in support of the mission and doing good science just for its own sake. Because NASA is by definition a mission agency, all its work is justified by its relationship to its missions. During reviews, panels are asked how each research area supports the agency's science goals.

1.5.1 "Moving toward uncertainty." There are difficulties in trying to evaluate projects by quantitative measures in light of how science is performed. The use of milestones, for example, implies a one-directional progression toward a goal. One who moves in this way is moving "toward certainty"—toward the proof of an expected conclusion—and scientists feel considerable pressure in their projects to "reduce uncertainty." Science, however, does not always develop in the expected direction, and the way to new understanding often means "moving toward uncertainty" in a project. For example, the discovery of the Antarctic ozone hole was disputed at first because the atmospheric models of the time did not predict it. The way toward discovery was toward uncertainty. The theorists had to go back and revise their models in the face of a fundamental advance. If the response to GPRA involves an excessive dependence on metrics, it could dissuade agencies from accepting uncertainty and moving toward new ideas.

1.5.2 A need for flexibility. Once a strategic goal is decided, there should be flexibility to move in new directions if the present direction proves unproductive. Such decisions should benefit from input from the scientific community.

1.5.3 The issue of control. Results of some programs might be out of NASA's control, such as educational, scientific, and commercial outreach. In educational outreach, for example, outcomes depend on the educational process itself. It should be assumed that good material will be used.

In the case of launch data, there is a concern that it will be decoupled from science in mission objectives because it is easy to quantify. There are other phases of the program, such as design review, that should be metrics, and their success should not depend on the actual launch, which is always subject to slippage. For example, the Terra launch slip led to six FY1999 targets not being met. Similarly, research partnerships should be evaluated in ways such that NASA's lack of control in partner-led areas does not unduly prejudice the results.

2. How is the result communicated to different audiences (e.g., S&T community, advisory committees, agency leadership, Administration, Congress)?

One of the goals of GPRA is to allow the various agencies and stakeholders to develop common nomenclature to deal with the evaluation of research. In addition, GPRA criteria should allow the agency to retain some flexibility and not place it in a straitjacket in its dealings with Congress and other oversight bodies. GPRA reports, in general, must be understandable to a wide array of people, but compliance requirements must recognize the need for technical discussion to capture the full reasoning behind the science.

2.1 Explaining the rationale for research.

GPRA documents have to explain to committees why particular things are done. One example is the goal of putting a spacecraft into an orbit 50 km above the surface of an asteroid. One could explain this goal by stating the simple metric that such a goal can allow photography of the asteroid's surface with 1-m resolution, but a qualitative rationale might be more appropriate. For example, a 50-km orbit is desirable because it lets us see the surface well enough to understand the internal process of the asteroid that influenced the formation of the surface.

2.2 Freedom to change course.

Congress might benefit from additional knowledge about the give-and-take of the scientific funding process. For example, it is common knowledge in the scientific community that principal investigators often change course from the plans they outline to their funding agencies. Such course changes are almost inevitable in pursuing the unknown. They do not indicate failure or willful disregard for the contract with a funding agency. Rather, the pursuit of a new direction is an indication of the "moving toward uncertainty" described above—the evolutionary process that leads to new knowledge.

In addition, NASA should take advantage of its strength in communication to better explain to the public what makes NASA unique, such as its effectiveness in interdisciplinary research and its ability to establish metrics for complex, interdisciplinary programs.

2.3 Communicating with the public.

Several participants congratulated NASA on the fullness and diversity of its communication with the public, including research publications, data archives, and Web sites. The agency has made efforts to increase the public's access to knowledge generated by NASA through exhibits, interviews, and news articles. It also assists in the location and retrieval of NASA-generated knowledge through help desks, image-exchange programs, and the Web site. Participants urged even more efforts like those to communicate the kinds of results sought by GPRA.

3. How is the result used in internal and external decision-making?

The desired result of the GPRA response is to make clearer to the public how government funds are being used to benefit the public. GPRA is also intended to be used by Congress to facilitate oversight activities and to make budget decisions, although GPRA has not yet been used for budgeting purposes.

3.1 The question of internal change.

The agency discussed at some length how these descriptions and their judgments can be used internally. At this stage in the evolution of the act, the GPRA "overlay" of NASA's extensive review mechanisms has not yet brought about new mechanisms for program change, although participants felt that there should be consequences.

3.2 Unhelpful comparisons.

Several participants expressed the concern that annual GPRA performance evaluations can lead to misunderstandings of the performance and value of long-term R&D. The present GPRA process generates expectations that "value-added outcomes" that benefit the American public should be reached each year by every research program. That is, a dollar of investment should earn at least a dollar of return, like a savings account. In fact, a research portfolio is more like the stock market, featuring many short-term ups and downs and the occasional long-term "home run." An effective GPRA reporting process would better communicate the high-risk, high-reward nature of research and provide convincing evidence of its value and continuing contributions to society.

NASA Focus Group Participant List
October 30, 2000

Panel Members:

Alan Schriesheim
Director Emeritus
Argonne National Laboratory
Argonne, Illinois

Wesley T. Huntress, Jr
Director, Geophysical Laboratory
Carnegie Institution of Washington
Washington, D.C.

Louis J. Lanzerotti
Distinguished Member of the Technical Staff
Bell Laboratories, Lucent Technologies
Murray Hill, New Jersey

Herbert H. Richardson
Associate Vice Chancellor of Engineering and Director, Texas Transportation Institute
The Texas A&M University System
College Station, Texas

Participants:

Nicholas Bigelow
Department of Physics and Astronomy
The University of Rochester
Rochester, New York

Erin C. Hatch
Analyst in Space and Technology Policy
Congressional Research Service
Library of Congress
Washington, D.C.

Allen Li
Director, Acquisition and Sourcing Management
US General Accounting Office
Washington, D.C.

Robin Nazzaro
Assistant Director
US General Accounting Office
Washington, D.C.

Agency Representatives:

Rich Beck
Director, Resource Analysis Division
NASA Headquarters
Washington D.C.

Ann Carlson
Assistant to the NASA Chief Scientist
NASA Headquarters
Washington, D.C.

Phil Decola
NASA Headquarters
Washington, D.C.

Jack Kaye
Director, Earth Sciences Research Division
NASA Headquarters
Washington D.C.

Guenter Riegler
Director, Space Sciences Research Program Management Division
NASA Headquarters
Washington, D.C.

Eugene Trinh
Director, Physical Sciences Research Division
NASA Headquarters
Washington, D.C.

SUMMARY OF THE NATIONAL SCIENCE FOUNDATION FOCUS GROUP

1.What methodology is used for evaluating research programs under GPRA?

1.1 Overview.

The National Science Foundation (NSF) is an independent agency of the US government, established by the NSF Organic Act of 1950 to "promote the progress of science; to advance the national health, prosperity and welfare; and to secure the national defense." NSF is governed by the National Science Board (24 part-time members) and a director, with a deputy director and assistant directors.

Its mission is unique among federal agencies in that it supports only extramural research, conducted primarily by principal investigators and groups in universities and other institutions. Other agencies, such as the Department of Defense, support large research programs, but the research components receive only a minor fraction of those agencies' budgets. At NSF, virtually the entire $4 billion budget is devoted to research (minus a portion—5% to 6%—spent to administer grants and awards).

1.1.1 Special challenges in complying with GPRA. Because of its unique charter, NSF faces special challenges in complying with GPRA. The first is that it has only limited control over the extramural research it funds. The agency relies on the advice of external, independent peer reviewers to evaluate the 30,000 applications received each year, of which about 10,000 are awarded.

A second challenge is that the progress and timing of research results are seldom predictable or measurable. Awardees may change the direction or emphasis of research projects as they discover new opportunities or knowledge.

Third, projects funded by NSF unfold over multiyear periods, and their results usually do not synchronize with the annual reporting requirements of GPRA. The agency has not found a way to provide annual evaluations of projects that have longer-term objectives.

A fourth challenge is a fundamental tension between the NSF organic mission and GPRA's requirement for quantitative metrics to evaluate research programs. Most research, especially basic research, is seldom measurable in quantitative ways. Potential impacts are difficult to predict and require long time frames. It is difficult to attribute specific causes and effects to final outcomes.

A fifth challenge is to comply with GPRA's objective of correlating performance goals and results with specific budgetary line items. Because of the timing of NSF grants and the progress of research, NSF cannot predict what the results of its programs will be in a given year.

Other challenges are finding a large number of experts who are qualified and independent, attributing success to a project that has multiple sources of support, and avoiding overconservative project selection that could inhibit the high-risk science that leads to high rewards.

To address those challenges, NSF has adopted an alternative reporting format, as approved by the Office of Management and Budget (OMB). The format relies on a mixture of quantitative and qualitative measures and relies primarily on expert review at the project and program levels. Within this format, specific research projects are monitored annually, and the progress of research programs is evaluated retrospectively every 3 years.

1.2 What level of unit is the focus of the evaluation?

For purposes of GPRA, NSF views its entire basic-research enterprise as a single "program." It has chosen this route in part because including a discussion of its individual research projects (some 10,000) or even individual research programs (about 200) is not practical. That is one reason, as several participants pointed out, why NSF's "results" cannot be matched with budgetary line items.

The core of the research enterprise is the individual research project; most of them are university-based. Some 95% of these projects are merit-reviewed before funding and then reviewed annually for progress by NSF staff. The merit-review (or expert-review) process continues under GPRA, although it does not appear specifically in GPRA reporting, because of the huge number of projects.

1.2.1 Outcome goals. The 200 agency-wide programs include directorate and cross-directorate programs, NSF-wide initiatives, small-business awards, the award program for individual investigators, and grants for small and large facilities. The activities of all those programs are included in evaluating broad outcome goals for the agency. For example, the FY1999 GPRA outcome goals listed by NSF are the following:

- Discoveries at and across the frontier of science and engineering.
- Connections between discoveries and their use in service to society.
- A diverse, globally oriented workforce of scientists and engineers.
- Improved achievement in mathematics and science skills needed by all Americans.
- Timely and relevant information on the national and international science and engineering enterprise.

For the first (and dominant) goal of "discoveries," NSF asked its reviewers to award one of two grades for FY1999: successful and minimally effective. Performance was to be judged "successful" when NSF awards led to important discoveries, new knowledge and techniques, and high-potential links across disciplinary boundaries. Performance was to be judged "minimally acceptable" when there was a "steady stream of outputs of good scientific quality." Officials found, however, that this combination of grades was not helpful to its reviewers, and for the FY2000 NSF has replaced these categories with: successful and not successful.

1.3 Who does the evaluation of the research program under GPRA?

NSF depends on two populations of reviewers to evaluate its programs. At the "grass roots" level, some 95% of individual research projects are approved and reviewed by independent expert reviewers (a small number are initiated internally by the director or others). These reviewers provide what an NSF representative called a "high-level front-end filter" for agency-supported research. Each project is reviewed annually and reviewed every 3 years for integrity and progress. This level of reviewing falls below the aggregation level of agency-wide GPRA reporting.

1.3.1 Committees of Visitors. At the much higher program level, NSF relies on its traditional external Committees of Visitors (COVs) to review integrity of process and quality of results of program portfolios every 3 years. These committees include people who represent a balanced array of disciplines, fields, and activities affected by outputs or outcomes of NSF-supported work and members of industry, government agencies, universities, foreign research communities, and other potential users. Each must be "credible" and "independent" (although independence is often difficult to judge in fields where many of the experts rely on a small number of funding sources and employers). Approximately 20 COVs meet each year to assess 30% of the NSF portfolio.

1.3.2 Advisory Committees. At the highest level are the directorate advisory committees, whose members are selected not only for expertise and perspective, but also for diversity of point of view. They are asked to review activities at the directorate and cross-directorate levels. Each advisory committee submits a report assessing the directorate each year.

For the GPRA report itself, NSF uses reports from each directorate's advisory committee. It combines those with COV reports (42 were submitted in 1999) to prepare an NSF-wide report. Each directorate may also use as input information for COVs and advisory committees, individual project reports (as examples or "nuggets" of high-quality research), budget justifications, press releases, annual-report materials, and National Research Council or other reports on the status of work supported by NSF.

1.4 What criteria are used for the evaluation?

In a broad sense, NSF relies on multiple criteria in evaluating programs. These include peer (expert) review of proposals, past performance of the principal investigator or group, community input, and input from the scientific community and public. Both qualitative and quantitative criteria may be used as tools.

1.4.1 Merit and impact. At the nitty-gritty level of proposal evaluation, reviewers are asked to look at two primary criteria:

- What is the intellectual merit of the proposed activity?
- What are the broader impacts of the proposed activity?

Out of this evaluation come two results. The first is advice as to whether to fund a proposal, and the second is a suggestion of the size of the award.

1.4.2 Process and results. The COVs are asked to evaluate both the *process* and the *results* of research programs. COVs provide NSF with expert judgments of the degree to which outputs and outcomes generated by awardees have contributed to the attainment of NSF's strategic goals. They also assess program operations and management.

The advisory committees are asked to review the reports of the COVs and to take a broader view of agency activities. Their reviews include specific and general guidance for program managers and are intended to influence practice at the managerial level.

An important criterion in evaluating any NSF program is the extent to which it promotes development of human resources. This criterion is stated in the NSF Act, which directs the agency to support "programs to strengthen scientific and engineering research potential." NSF has goals to promote the development of human resources within the agency and in the scientific community.

1.4.3 Quality, relevance, and leadership. As suggested in the original report of the Committee on Science, Engineering, and Public Policy (COSEPUP) on GPRA, NSF uses the criteria of quality and relevance in its evaluations. It focuses less attention on the third criterion, leadership, although its expert reviewers often take leadership status into account. They have not, however, found a way to assess leadership through international benchmarking, as proposed by COSEPUP.

1.4.4 A mix of qualitative and quantitative means. COSEPUP also addressed the issue of whether to rely more heavily on qualitative or quantitative means to evaluate research. It suggested that basic research can best be evaluated by expert review, which makes use of quantitative measures wherever appropriate. NSF uses a mix of both, depending on the material being reviewed. For the outcome goals and results of research, qualitative measures are used. For "investment process goals," a mixture of qualitative and quantitative means are used. For management goals, quantitative means predominate.

1.5 How does the selection and evaluation of projects relate to the evaluation of the research program?

Because NSF evaluates its "research program" on an agency-wide basis, the evaluation for the purpose of complying with GPRA is not directly related to the selection and evaluation of individual projects. As suggested above, the 10,000 or so projects selected each year cannot be discussed individually in any meaningful way for a single report.

1.5.1 Differences of scale. At the same time, the selection and evaluation of projects do form the heart of NSF's activities, and the nature of research is central to everything it does. Yet, OMB does not expect each project to be evaluated under GPRA. The agency-wide evaluation is performed on a different scale from a single-project evaluation, but by the same principles. Hence, NSF's GPRA performance plan for FY2001 includes the statements that "even the best proposals are guesses about the future," "true research impact is always judged in retrospect," and "the greatest impacts are often 'non-linear' or unexpected."

In contrast, some of NSF's projects are easily quantifiable and are evaluated on that basis. For example, the Laser Interferometer Gravity-Wave Observatory near Hanford, Washington, is the agency's largest investment. It was delivered on time and under budget, and the program as a whole was judged a success from that perspective. But from a research perspective, the agency cannot yet know whether the observatory will detect gravitational radiation and bring new knowledge to the world.

1.5.2 An important disconnect. In one important respect, the selection of projects is disconnected from the evaluation of the research program. NSF advisory committees must submit information for its assessment in September, before the books close on the current year.

Similarly, the agency is about to begin work on the performance plan for 2002, but it does not yet have the report for 2000 to know where it should be making changes for 2002.

An NSF representative said, "In addition, it may take a year or two to put a solicitation out, receive proposals, evaluate them, and make awards. Because most awards cover five years, the time lag between putting out a request for proposals and meaningful results may be six or seven years (or more)."

2. How is the result communicated to different audiences (e.g., S&T community, advisory committees, agency leadership, Administration, Congress)?

The issue of "transparency" in GPRA reporting was discussed at length. On the one hand, NSF has received high praise for communicating openly with its many stakeholders, including Congress, OMB, the Office of Science and Technology Policy, NSF's National Science Board, NSF advisory committees, the National Academies, S&E professional societies, the academic community, and the general public (partly through the NSF Web site).

2.1 The issue of COV reports.

The reports of the COVs are not readily available. Several participants urged easier access to COV reports or perhaps summaries of them that could be posted on the Web.

2.2 Communicating with Congress.

There was considerable discussion about how NSF could better describe to Congress how it judges good science. Legislators want more transparency, worrying that they are being asked to accept scientific judgments without knowing how those judgments work. However, communicating with Congress is difficult because many staffers are political appointees without scientific training. NSF should take the initiative by telling its story better, educating the staff where needed, and showing the value of basic research in ways that are useful for Congress.

2.3 The risk of self-serving reports.

Participants discussed a shift in the attitude of principal investigators caused by NSF changes, notably the "fast lane" application mechanism and the speed of electronic filing. In the past, principal investigators would spend as little time as possible on NSF reports. In the last year, that has changed, and people are trying to get their reports to NSF for the sake of greater visibility for their projects. This, in turn, brings the danger of self-serving reports and promotion of one's research agenda.

3. How is the result used in internal and external decision-making?

One consequence of complying with the GPRA performance plan has been to simplify NSF goals. Five broad agency goals have been whittled down to three:

- Ideas (discoveries, learning, and innovation).
- People (the workforce).
- Tools (physical infrastructure for research and education).

The previous five goals placed insufficient emphasis on equipment and facilities, especially new information resources, and did not match well with the reporting requirements of GPRA.

3.1 Some internal benefits

- The process helps the agency to focus on issues that need attention, such as (in FY1999) the fast-lane area and the use of merit review criteria.

- It has helped to improve management efficiency and effectiveness. Specifically, it has helped to collect information, focus activities, sharpen the vision of the directorate, and see in a broad way what the agency is doing.
- It puts more discipline into planning. Before, the agency had a "visionary plan," but it was not very well connected to implementation. Now, NSF has to connect that strategic plan all the way down to program level and assessment.
- It helps to increase accountability, and for this the help of the scientific community is needed. The NSF Accountability Report for FY1999 received the highest marks for a government agency.

3.2 A time-consuming process.

One important result, which stimulated considerable discussion, was the amount of time and effort the agency has devoted to compliance. The requirements for documentation are increasing. Principal investigators are being asked for more, COVs are asked to digest more, and advisory committees have substantially more to do, as does everyone at the directorate level.

NSF Focus Group Participant List
November 21, 2000

Panel Members:

Enriqueta Bond
President
The Burroughs Wellcome Fund
Research Triangle Park, North Carolina

Alan Schriesheim
Director Emeritus
Argonne National Laboratory
Argonne, Illinois

Rudolph A. Marcus
Arthur Amos Noyes Professor of Chemistry
California Institute of Technology
Pasadena, California

Stuart A. Rice
Frank P. Hixon Distinguished Service Professor
James Franck Institute
The University of Chicago
Chicago, Illinois

Max D. Summers
Professor of Entomology
Texas A&M University
College Station, Texas

Bailus Walker, Jr.
Professor of Environmental and Occupational Medicine
Howard University
Washington, D.C.

Participants:

David Ellis
EHR Advisory Committee Chair-NSF
President and Director, Museum of Science
Boston, Massachusetts

Irwin Feller
SBE Advisory Committee Chair-NSF
Institute for Policy Research and Evaluation
Pennsylvania State University
University Park, Pennsylvania

Eric A. Fischer
Senior Specialist in Science and Technology
Congressional Research Service
Library of Congress
Washington, D.C.

Robert Hull (by conference call)
Department of Materials Science and Engineering
University of Virginia
Charlottesville, Virginia

Genevieve Knezo
Specialist, Science and Technology Policy
Congressional Research Service
Library of Congress
Washington, D.C.

Robin Nazzaro
Assistant Director
US General Accounting Office
Washington, D.C.

David Radzanowski
Program Examiner, Science and Space Programs Branch
Office of Management and Budget
Washington, D.C.

Diane Raynes
US General Accounting Office
Washington, D.C.

Jean'ne Shreeve
Department of Chemistry
University of Idaho
Moscow, Idaho

Jack Sandweiss
Donner Professor, Department of Physics
Yale University
New Haven, Connecticut

David Simpson
The IRIS Consortium
Washington, D.C.

Agency Representatives:

Robert Eisenstein
Assistant Director for Math and Physical Sciences
The National Science Foundation
Arlington, Virginia

Paul Herer
Senior Staff Associate, Office of Integrative Activities
The National Science Foundation
Arlington, Virginia

Lorretta Hopkins
Staff Associate
Office of Integrative Activities
The National Science Foundation
Arlington, Virginia

Nathaniel Pitts
Director, Office of Integrated Activities
The National Science Foundation
Arlington, Virginia

Martha Rubenstein
Budget Division Director
The National Science Foundation
Arlington, Virginia

SUMMARY OF THE DEPARTMENT OF ENERGY FOCUS GROUP

1. What methodology is used for evaluating research programs under GPRA?

1.1 Overview.

1.1.1 Management structure. For management purposes, the Department of Energy (DOE), an $18.9 billion agency, is divided into four "business lines," including science, national security, energy, and environment. Most S&E research is supported by the Office of Science, whose five sub-offices are budgeted at just over $3 billion for FY2001. The $1 billion Office of Basic Energy Sciences (BES), was the one most extensively discussed at the workshop.[1]

About half the budget of the Office of Science (SC) is allocated to research and divided 60/40 between research at its laboratories and research at universities. BES alone funds about 3,500 grants in 250 colleges and universities throughout the United States. Large laboratories and user facilities (26 major facilities and 12 collaborative research centers) receive over 30% of the office's budget; smaller portions go to major construction projects (currently featuring the Spallation Neutron Source and a high-energy physics project to study neutrino oscillations), capital equipment, and

[1]The other four science suboffices are in biologic and environmental research, high-energy and nuclear physics, fusion energy sciences, and advanced scientific computing research.

program direction. The laboratories are shared by many users from academe, government, and industry. Most laboratories, such as Brookhaven National Laboratory on Long Island, are government-owned and contractor-operated (GOCO). The contractors may be universities, not-for-profit organizations, industries, or consortia.

1.1.2 A shortage of needed information. The scientific offices within DOE have found it difficult to comply with GPRA. The agency as a whole lacks a good system for tracking data that it needs to report on all its activities. The agency attempted to rectify this situation through a substantial investment in a federal government R&D database, but the lesson of that experience was that the agency needed its own system.

DOE tried at first to use a systemwide framework that emphasized the agency's financial structure in the hope that it would be easy to reconcile with the budget. This financial overlap, however, did not accurately represent what the agency does, and it was divorced from actual planning. The linkages between this plan and what occurs at the program-manager level were weak, and the plan did not match well with GPRA. The General Accounting Office (GAO) was critical of the process, and administrators felt both external and internal pressure to change it.

1.1.3 A new planning model. Planners knew that a new planning model would have to be flexible because each new administration formulates different policies. But GPRA requires uniformity and a clear linkage between the performance plan and the performance report. The model would have to be able to show how the actions of DOE administrators result in excellent science at universities.

As a result, SC is currently attempting to design a new strategic planning process to characterize the research it is doing and link its GPRA reports more logically to science.

1.1.4 "Science is different." The reason for this attempt is that scientific organizations are different from other agencies because scientific research is different from other activities. Therefore, strategic planning for science should also be different.

Through a literature survey and focus groups, the agency is trying to develop a more "holistic view of the pathways that lead to excellent science." The goal is to describe the pathways that an organization should take to achieve excellent science.

The agency has been studying this subject for one year and has now described an environment that "fosters excellent research at the performing institution." A suggested framework includes two dimensions (internal focus and integration, and external focus and differentiation) and four perspectives of effective organizations: human-resource development, internal support structures, innovation and cross-fertilization, and setting and achievement of relevant goals.

1.2 What level of unit is the focus of the evaluation?

The agency has had difficulties in setting an appropriate level of unit for evaluation and in finding relevant performance measures. The individual programs had no early basis for deciding what level of aggregation to use or how many measures to apply. Therefore, some program-level reports have been very detailed and others more general, depending on the approach of individual program directors.

1.2.1 Reviewing methods. Below the level of programs (and of the GPRA performance reports), much of DOE's research budget is allocated to support individual, investigator-driven research projects in universities. These projects are evaluated individually by traditional peer review—that is, the same external, independent review system used by the National Science Foundation, the National Institutes of Health, and other agencies that support external research.

For research supported and overseen directly by the agency, the unit of evaluation is usually the laboratory or the program within the laboratory. These units have long-established means of evaluation through external and program reviews that have been maintained for GPRA.

Some subtlety is involved in evaluating large facilities during construction or operation. Most of them, such as Spallation Neutron Source, are "one-of-a-kind" projects whose very construction may involve cutting-edge science. Once they are operational, the "maintenance" expenses for such facilities may become difficult to distinguish from the "research" expenses for the purpose of GPRA.

The agency also measures its contribution to S&E human resources. The agency maintains a commitment to supporting graduate and postdoctoral education; despite budget losses in the laboratories, it has roughly maintained existing levels of grants to universities.

1.3 Who does the evaluation of the research program under GPRA?

1.3.1 Peer reviewers. For the university grant program, virtually all individual projects are evaluated by regular peer review under the Office of Science's Merit Review System guidelines. This external process conforms to standard independent peer-review procedures.

For laboratory research programs and facilities (e.g., Argonne National Laboratory and the Princeton Plasma Physics Laboratory), a formal documentation system similar to peer review is the norm. For example, BES evaluates the research projects it funds according to procedures described in *Merit Review Procedures for Basic Energy Sciences Projects at the Department of Energy Laboratories*. These procedures are patterned after those given for the university grant program. Peer review at the laboratories is intended to provide an independent assessment of the scientific or technical merit of the research by peers who have "knowledge and expertise equal to that of the researchers whose work they review."

1.3.2 Technical experts. Major construction projects are evaluated by technical experts who look at relatively straightforward criteria, including cost, schedule, technical scope, and management ("Lehman reviews"). Reviews of major projects are typically held twice per year and may include 30-40 independent technical experts divided into six to eight subpanels.

1.3.3 Advisory committees. For each of the five SC programs, the evaluation procedure also includes advisory committees. For example, the 26-member Basic Energy Sciences Advisory Committee (BESAC) meets two to four times per year to review the BES program, advise on long-range planning and priorities, and advise on appropriate levels of funding and other issues of concern to the agency. BESAC subcommittees focus on more specific topics, such as neutron-source upgrades and DOE synchrotron radiation sources. Users of BES facilities are surveyed annually and asked for quantitative information about publications, patents, Cooperative Research and Development Agreements, prizes and awards, and other achievements.

BESAC reviews do not feed directly into the GPRA process. The committee looks at peer reviews, contractor reviews, citation indexes, major awards, and any other relevant information; distills the information; and reports directly to the director of the Office of Science. The committee attempts to clarify why DOE is supporting particular programs and to gauge the contribution of individual facilities to the agency's research effort.

1.3.4 Dual reviews for GOCOs. GOCOs are assessed by both DOE and the contractors. The agency does not rely on a contractor's review, because the contractor has an incentive to provide a favorable review to justify its compensation. Instead, the agency does annual "contractor appraisals" by using independent peer review. Ratings are given for

- Research quality.
- Relevance to mission.
- Research facilities.
- Research management.

Overall appraisals are "rolled up" from individual laboratory reviews for all programs. These contractor appraisals affect performance fees and contract renewals.

1.4 What criteria are used for the evaluation?

DOE's budget narrative system lists a summary of "budget guidance" items, beginning with program mission, program goal, and program objectives. DOE is attempting to conform GPRA's requirements with the budgetary requirements.

1.4.1 Separating excellence from relevance. The new system departs from the intent of the three COSEPUP criteria, however, by yoking the first two, excellence and relevance. These measures should be separated. Some excellent research may not be relevant to the agency's mission, and some relevant research may not be of excellent quality.

1.4.2 150 measures. SC has been using more than 150 performance measures, which DOE representatives (and GAO) acknowledge is an unwieldy number. This system has not been helpful in assessments to date, partly because the measures are not specific enough, do not clarify the DOE role, do not include means of validation and verification, and do not have clear links to the DOE strategic plan and budget.

The agency's "emerging measures" are patterned more closely on the COSEPUP recommendations by including leadership. To measure the level of leadership, the agency is contemplating the use of the "virtual congress," as suggested in the COSEPUP report.

1.4.3 Studying new criteria. The new criteria for performance metrics—now being studied by a group led by Irwin Feller, of Pennsylvania State University—are being examined in the hope of allowing a response to GPRA that is "grounded in research." The criteria will attempt to include the following elements:

- Reasonable metrics (that is, reasonable for assessing a science agency).
- Excellence in science management (a 3-year study that benchmarks best management practices was launched in January 2000).
- Science "foresighting" (another 3-year study is examining science trends "out to 25 years").
- Portfolio analysis (using information-technology tools, including deep data-mining, to characterize the research portfolios of the Office of Science, the federal government, and the "international S&T research portfolio").
- Miscellaneous efforts (to apply organizational and management theory).

1.4.4 The need to take risks. The Office of Science also uses the criterion of risk in evaluating its programs. Without taking risks in research, programs and projects are unlikely to achieve the high-reward payoffs of the best investigations. Missions need flexibility in choosing research directions because peer review by itself is inherently conservative.

1.5 How does the selection and evaluation of projects relate to the evaluation of the research program?

Participants discussed at some length the "charter" of DOE and how DOE managers decide to include or exclude various programs or research topics from this charter. This issue is important in assessing the relevance of research for GPRA.

1.5.1 Complexities of project selection. The process of selecting projects is complex and combines information from the Office of Strategic Planning, input from advisory committees, and program decisions made internally. The users of DOE facilities come from many institutions, with many agendas, and DOE does not want to restrict the scope of research for those who are using the facilities in productive ways.

2. How is the result communicated to different audiences (e.g., S&T community, advisory committees, agency leadership, Administration, Congress)?

In its report to Congress on the usefulness of agency performance plans, GAO noted that SC's FY2000 plan was "moderately improved" over the FY1999 plan but still bore little relationship to budgeting. The agency felt that more improvement was needed and for the succeeding year attempted to follow the structure of the budget more closely. Therefore, it organized the performance goals by budget accounts and annotated the performance goals with linkages to the strategic plan by identifying the strategic objectives they support.

2.1 Meeting with oversight staff.

The agency also met with congressional staff and agreed to characterize its results by four categories: exceeded goal, met goal, nearly met goal, and below expectation. Each rank was based on deviation from the expectation established in the performance goal.

This was done in response to GAO's concern that baselines and context had not been provided to compare with performance.

The agency has also added a section on verification and validation under each decision unit, including periodic guidance, reviews, certifications, and audits. Because of the size and diversity of the department's portfolio, verification is supported by extensive automated systems, external expert analysis, and management reviews.

2.2 Communicating about the new model.

There is considerable communication between DOE and GAO. After receiving a GAO report indicating that procedures for peer review vary among federal agencies, the House Science Committee asked GAO to investigate. GAO randomly sampled 100 BES research projects and concluded that the agency was performing merit review properly and following the established procedures.

3. How is the result used in internal and external decision-making?

3.1 GPRA results do not yet influence funding.

A common assumption about GPRA is that its results will be used to make funding decisions. However, many congressional staffs have not yet found ways to match performance results with funding decisions, because the process is still new and results are not often easily aligned with budgetary structure.

3.2 A critique of GPRA reports.

Performance metrics do little good unless they embrace the scientific effort as a whole. For example, metrics of construction projects say little about the value of the science that they are intended to support. It is important to use quality, relevance, and leadership as evaluation criteria; the agency should not try to review the whole portfolio every year.

Office of Science officials stated that they are suggesting a process very similar to this.

3.3 One result is DOE's new model.

Indeed, one result of DOE officials' attempts to evaluate their scientific research for GPRA has been to convince the agency of the desirability of the new assessment model that they are studying. The goals of the study are to

- Investigate how funding agencies can foster excellent science.
- Focus on the impacts of interactions among the Office of Science and science-performing organizations.
- Identify relevant research in organizational effectiveness and science management.
- Fill gaps in knowledge or public-sector issues in management of scientific research.
- Formulate strategies for dealing with large changes in research and funding environments.

Preliminary results have been mentioned above, but much of the study remains to be accomplished.

The agency noted that its reviews do have results—that a poor review of the construction of the Spallation Neutron Source had resulted in substantial changes in senior management.

DOE Focus Group Participant List
November 29, 2000

Panel Members:

Alan Schriesheim (Cochair)
Director Emeritus
Argonne National Laboratory
Argonne, Illinois

Morris Tanenbaum
Retired Vice Chairman and Chief Financial Officer, AT&T
Short Hills, New Jersey

Participants:

Eugene W. Bierly
Senior Scientist
American Geophysical Union
Washington, D.C.

Jack E. Crow
Director, National High Magnetic Field Laboratory
Florida State University
Tallahassee, Florida

Eric A. Fischer
Senior Specialist in Science and Technology
Congressional Research Service
Library of Congress
Washington, D.C.

Richard D. Hazeltine
Professor of Physics
Director, Institute for Fusion Studies
University of Texas at Austin
Austin, Texas

Michael Holland
Program Examiner
Office of Management and Budget
Washington, D.C.

Genevieve Knezo
Specialist, Science and Technology Policy
Congressional Research Service
Library of Congress
Washington, D.C.

Robin Nazzaro
Assistant Director
US General Accounting Office
Washington, D.C.

Fred Sissine
Specialist in Energy Policy
Congressional Research Service
Library of Congress
Washington, D.C.

David Trinkle
Program Examiner, Science and Space Programs Branch
Office of Management and Budget
Washington, D.C.

Agency Representatives:

Patricia Dehmer
Associate Director, Office of Basic Energy Sciences
US Department of Energy
Germantown, Maryland

William J. Valdez
Director of the Office of Planning and Analysis
US Department of Energy
Washington, D.C.

SUMMARY OF WORKSHOP

On December 18-19, 2000, the Committee on Science, Engineering, and Public Policy (COSEPUP) sponsored a 2-day workshop on the Government Performance and Results Act (GPRA). The purpose of the workshop was to allow participants to summarize the points raised in five agency-specific focus groups[1] held over the previous 3 months, to review these points with representatives of the agencies and federal oversight groups, and to formulate their own conclusions and recommendations. This document summarizes the main points discussed at the workshop.

This summary refers several times to the first GPRA report by COSEPUP.[2] The executive summary of that report is included as Appendix E. It also reiterates the findings of the first report, with emphasis on the first four recommendations: research programs, including basic research, should be evaluated regularly; the methodology of evaluation should match the character of the research; the primary method for evaluating research programs should be expert review; and agencies should describe in their GPRA plans and reports the goal of developing human resources.

[1]See Appendix C for summaries of focus groups with Department of Defense, Department of Energy, National Aeronautics and Space Administration, National Institutes of Health, and National Science Foundation.

[2]COSEPUP, *Evaluating Federal Research Programs: Research and the Government Performance and Results Act*, Washington, D.C.: National Academy Press, 1999.

Evaluating Basic Research

The language of GPRA strongly encourages agencies to evaluate all their activities, including basic research, with quantitative metrics that can be applied annually. Much of the research in the large mission agencies, such as the Department of Energy (DOE) and the Department of Defense (DOD), is applied or developmental research, which is more amenable to quantitative measurement. But the panel heard from agency representatives that they could not find useful quantitative metrics to evaluate the results of basic research.

Limits of Quantitative Metrics

It is true that quantitative measures are used to evaluate researchers, research proposals, and research programs throughout science. Some of these measures are the number of publications, the number of times papers are cited by others, the number of invited talks given, and the number of prizes won. There was universal agreement, however, that the usefulness of such quantitative measures by themselves is limited. A citation index, for example, is a relatively crude measure in that it does not measure the originality of papers, the quality of publications, the number of co-authors, or other qualitative conditions that are fundamental to understanding their value. Many researchers publish large numbers of papers, each of which represents only a slight variation on the preceding one. Similarly, junior researchers who belong to very large research groups might play almost no role in the design of an experiment whose report they help to write. On the other side of the argument, "routine" papers might have greater value than first supposed. For example, a simple paper on methodology might contain an original insight into some technique that proves to be of great value. Discriminations of those kinds are best made by experts who are asked specifically to focus on the work of a particular person or laboratory.

The Value of Expert Review

The judgment of experts as a form of "measurement" has true value because of the reviewers' deep knowledge of a field and of the people who work in it.

Several basic points were identified:

- Because basic research is an open and free inquiry into the workings of nature, the eventual significance or utility ("outcome") of basic research cannot be predicted.
- For purposes of evaluation, one can evaluate basic research on the basis of whether it is producing high-quality knowledge ("output") that is relevant to the mission of the agency supporting the work.

Panelists offered several illustrations of the difficulty of trying to evaluate basic research annually with quantitative metrics.

- Quantitative evaluation can stifle the very inquiry it is trying to measure. For example, a researcher sets a measurable goal at the beginning of the year. In July, the researcher discovers a more promising direction and decides to alter course. The original goal is no longer meaningful. Even though the change of direction benefits the inquiry and the research program in the long run, the researcher would receive a low "GPRA grade" for the year on the basis of the original quantitative goal.
- The original language of GPRA encourages agencies to design their budgets in a way that links all expenditures with defined goals for that budget year. If an agency programs all its money at the outset of a budget cycle, it cannot move in a new direction when the promise of that direction is revealed.

Some agency experiences

A representative of the Environmental Protection Agency (EPA) described his agency's struggle to conform to GPRA. Some of the basic research supported by the agency, he said, does not easily align with goals expressed in the budget, because the results are unknowable or because the agency cannot show results within the budget year.

A representative of the National Aeronautics and Space Administration (NASA) said that neither the Office of Management and Budget nor some of his own agency people seemed to appreciate the differences between NASA's basic research and other ways that the NASA mission is implemented. "All of what we do is not the same," she said, "but they expect that our plans and reports are going to look the same."

A representative of DOE noted that as a manager he was not in a position to directly judge the science supported by the agency. "We are science-managing, not science-performing. There is a tenuous link between what we do in the office and the actual performance of science at a laboratory or university." He said that science is performed according to its own standards of integrity and that the best science-management practices are those that will not have an adverse effect on science itself.

Several scientists expressed surprise that they had to explain the process of basic research each time new members came to Congress. They suggested that agencies coordinate presentations to communicate about this issue with oversight bodies better. They also noted that once oversight groups recognized the basic principles of science, they might understand better why micromanaging of agency research programs does not necessarily lead to better science.

Criteria for Evaluation

In its first report on GPRA, COSEPUP recommended the use of the following criteria to evaluate research programs: the *quality* of the research, the *relevance* of the research to the agency's mission, and *leadership*—that is, the level of the work being performed compared with the level of the finest research done in the same field anywhere in the world. The criteria were discussed frequently throughout the workshop.

Quality

Agencies have all incorporated excellence into their evaluations. Individual research projects are evaluated for their quality by panels of independent experts in the same field of research. At a higher level, research programs are usually evaluated by panels of independent experts.

However, there is considerable variation among agencies in how quality is assessed. The entire mission of the National Science Foundation (NSF) is research; a large, traditional structure of volunteer peer reviewers spends large amounts of time and effort in reviewing grant applications and grant renewals. In DOD, a very small proportion (1.5%) of the overall budget is dedicated to research, and this portion is peer-reviewed in the same manner as NSF-funded research by expert peers. Outcomes of most of DOD's work, however, which includes extensive programs of weapons testing, are more predictable, and the research component of the work is small. Instead of traditional peer review, the agency more often evaluates such work by marking its progress against established benchmarks. Because of such variations, the panel acknowledged that there are multiple approaches for gathering information on quality and for setting priorities and that agencies should be free to design their own approaches.

There was also acknowledgment of how much work is asked of the science and engineering communities in serving on review panels. Both NSF and DOD, in particular, as well as the National Institutes of Health (NIH), are sometimes accused of reviewing too much and of overtaxing their reviewers.

Relevance

Panelists concluded that agencies generally have methods for gauging the relevance of their research to their missions. These methods, however, vary widely among agencies and are seldom made clear in GPRA plans or reports.

Two views of relevance were discussed. The first is relevance as perceived by agency managers, who must decide what kinds of research are relevant to their mission objectives. The second is the view of the "users" of research. For example, the users of results of NIH research include pharmaceutical companies, hospital administrators, applied researchers, and doctors. NIH was asked whether it heard from such users, and a representative responded that the agency hears from them through its national advisory councils, which include scientists, health-care providers, and members of the public. The agency also holds workshops to gather general feedback.

The example of the Army Research Laboratory (ARL) was discussed. ARL uses both external peer groups and user groups to evaluate its research.

Peer committees are specifically designed to evaluate quality, but they are less well equipped to evaluate relevance. For that, expert committees must be augmented by members of the user community or experts in related fields.

Relevance was conceded to be easier to assess for entities like ARL, where researchers work closely with those who will use the outcomes of research. It might be more difficult to describe for NSF and NIH, where most research is performed externally, users might be unknown, and most research is basic research.

A panelist remarked that in assessing DOE research, users do have input, but it is seldom revealed in plans or reports.

Leadership

International benchmarking is the use of expert panels that include reviewers from both the United States and other countries to evaluate the leadership status of a country in a given research field. The goal of international benchmarking is to judge the "leadership level" of a program with respect to the world standard of research in that field. COSEPUP had written earlier that for the sake of the nation's well-being, the United States should be among the leaders in all major fields of science and be preeminent in selected fields of national importance.[3]

It was agreed that the agencies that focus on basic research, notably NSF and NIH, address the issue of leadership at least tacitly by funding their researchers competitively. By funding the best researchers, they are supporting the careers of the best scientists. But the leadership issue might be addressed more explicitly by including more foreign researchers on review panels. The discussants encouraged agencies to experiment with ways to increase the international perspective in their evaluation procedures.

One panelist cited an earlier COSEPUP experiment with international benchmarking, in which the United States was deemed to be the overall world leader in materials science and engineering, but the study revealed some fields in which the United States was not ahead. "If I were sitting in an agency", said the panelist, "that would worry me. The only way to get at this picture is through an international viewpoint."

DOE and several other agencies mentioned plans to experiment with international benchmarking. One agency representative cautioned that setting up such a program might take more time than is allowed in the framework of GPRA.

Education

The panel agreed that every agency that supports research has an interest in enhancing the education of graduate students, postdoctoral scientists, and active scientists. At the workshop,

[3]The rationale for these complementary goals is that the nation must be performing research at the forefront of a field if it is to understand, appropriate, and capitalize on current advances in that field, no matter where in the world they occur. Cite Goals report.

however, agency representatives seldom mentioned education in their presentations. The first GPRA report explicitly recommended the use of education as an evaluation criterion for purposes of GPRA compliance, and the panel reiterated this recommendation. Specifically, the panel recommended that the expansion or contraction of programs be assessed for effect on present and future workforce needs.

Aggregation of Research Programs for Purposes of Evaluation

Agencies support hundreds or thousands of individual research projects and they cannot evaluate all of them for the purpose of GPRA compliance. Therefore, they aggregate their projects to a large extent. Some agencies, such as the DOD, aggregate up to the program level; others, such as NSF, aggregate virtually all their projects into a single "research portfolio."

The topic of aggregation provoked extensive discussion, largely because a very high level of aggregation prevents insight into the evaluation of specific programs or divisions within programs. One criticism of high aggregation is that it is opaque to oversight bodies and others who want to understand how an agency makes decisions. An opposing view was that it is not appropriate for oversight bodies to "micromanage" agencies' selection and evaluation of individual programs or projects.

A workshop participant noted that some committees do not use GPRA documents when research activities are too highly aggregated. She noted that the law requires research activities to be described at the program and financing budget levels. In NSF GPRA documents, she said, the existence of specific programs or disciplines is not apparent, and it is not possible to weigh activities against goals. "Congress would like to know what you were trying to do. It would like clearer statements of objectives and accomplishments."

Other participants indicated that it was risky to try and that agencies are often caught between conflicting desires, because some committees want to see a high degree of detail and others do not.

A number of participants indicated that the degree of aggregation should be left up to the individual agencies. It was suggested, for example, that aggregation was easier in a mission agency, such as DOD, because its programs are more focused on specific goals, whereas NSF supports virtually all forms of research, which might not have predictable goals.

Some also said that when agencies choose a high level of aggregation they should also make clear how decisions are made below that level and provide access to materials that demonstrate the decision-making. Even though not all committees will want to read through the long, highly detailed documents used by agencies for internal planning, these documents should be available.

Another point made was that there has always been a tension between Congress and the scientific community about what kinds of research to pursue. In some cases, Congress would like to micromanage an agency's portfolio to pursue political or other nonscientific goals. In such cases, it is understandable that agencies prefer to shield their activities from decisions that can alter their program plans.

One Size Does Not Fit All

Another issue addressed at the workshop was the concept that "one size does not fit all." One of the most striking examples of difference can be seen in the research supported by two agencies: NSF and DOD. Nearly all NSF's budget is spent on research, but only about 1.5% of DOD's budget is. And yet each has to respond to the same GPRA requirements, even though the overall DOD GPRA plan barely has space to mention research at all, let alone deliver a detailed analysis of planning and evaluation methods.

An NSF representative said, "Our number one principle is to do no harm. One size doesn't fit all. If the shoe doesn't fit, it isn't your shoe."

A NASA representative explained the differences particular to her agency. At NASA, research activities are integrated across so-called enterprises, which are the major agency divisions. Each enterprise has a portion of a kind of research, and that portion must be integrated with the other activities of the enterprise, such as building hardware and planning space missions. It is difficult to explain the different qualities of scientific research within a GPRA document that comprehends an entire enterprise with all its diverse activities and goals.

Another agency representative expanded on the difficulty of the large mission agencies. Because most of their activities are not research, the agencies themselves might not emphasize or even understand the research process. One representative pointed out that GPRA reporting in his agency is done through the chief financial officer.

The Usefulness of GPRA

The panel asked many questions about the utility of GPRA for agencies: What benefits, if any, does it bring to your agency?

Some agencies saw benefit in being forced to examine management procedures more closely and to think in more detail about how their research activities served the objectives described in their budgets.

Other agencies were still struggling to make sense of the GPRA requirements and to fit them to their agency's structure and function. For example, EPA expressed a "lot of dilemmas." It felt a split between its overall mission and many of the science programs that supported that mission. Some of the programs supported basic research and could not be described annually in terms of outcomes—and yet both the oversight groups and agency administrators asked for such outcomes. Similarly, the US Department of Agriculture (USDA) described itself as "very mission-driven" but having core agencies that perform research. The representative felt that the most useful way to use GPRA was to apply it to program management, not to the research itself.

DOE expressed the most profound difficulties. A representative said, "We're required by different people to meet different requirements not of our choosing. We're trying to come to grips with GPRA by focusing on budget—adopting a planning process that allows us to embed performance goals in the budgeting process that makes sense from the GPRA point of view. Now we have gotten instructions from appropriators to strip out all high-level goals and instead to use performance measures in line items as statement of what we're trying to accomplish—$2-3 million items, very specific. This is a big problem."

DOD said that GPRA has not added value to its evaluation process, because the agency is using the same procedures that it did before GPRA. It still evaluates the quality and relevance of research with a GPRA-like process.

Some participants indicated that agencies do not appreciate the flexibility built into GPRA. That is, the law permits agencies to devise "alternative forms" of planning and evaluation when annual quantitative techniques are not appropriate. But some agencies that are expressing the most difficulty have not fully done so.

GPRA and the Workload of Agencies

The law does not allow agencies to hire additional staff or consultants to comply with GPRA, and its intent is not to impose an additional workload. But agency representatives described a considerable amount of extra workload in the form of meetings, workshops, and other activities. One representative offered an unofficial estimate that one-fourth to one-third of the time of some middle- and high-level officials was devoted to GPRA compliance.

Some workshop participants also expressed concern over the amount of time devoted to GPRA. They felt strongly that it should not replace or interfere with how agencies do their strategic planning. But they were optimistic that once agencies moved farther along the learning curve, they would be able to integrate GPRA reporting procedures with internal agency procedures in ways that benefit both but do not require additional time.

An NIH representative said that NIH had not had to change its internal planning or reporting procedures but felt that GPRA required special attention. She noted that GPRA work takes place in the context of other activities: planning, priority-setting, and producing other documents for 23 institutes and centers. For NIH as a whole, there are 55 strategic targets, only five of which had been discussed at the workshop. The rest—including training facilities, administration, grants, technology transfer, and priority-setting—are equally important parts of GPRA.

An NSF representative said that GPRA "is expensive for our agency." He said that the CFO, chief information officer, and many others all meet weekly to talk about it. The agency had to develop data systems to accumulate information for GPRA. It also affects the committees of visitors (COVs) that review NSF programs. The COVs used to study only the process of making awards and ensure that it was fair and honest. Since passage of GPRA, the agency has expanded the mission of COVs to evaluate the research results of past investments.

DOD indicated that it had been able to integrate GPRA into its processes. The panelists noted that the extra effort would probably decrease as agencies developed systems that responded to GPRA more easily. They also urged oversight bodies to help agencies to develop reporting formats that minimize the extra effort required.

Two issues of timing

Linking performance plans with the budget cycle

Most agency representatives reported difficulties in complying with the timing of GPRA requirements. They are required to send in their performance plans and performance reports with the annual budget. However, because budgets are due at the beginning of each year, sending in an annual performance plan with the budget requires preparation of the report before the year has ended.

NSF explained its difficulties this way: "If you haven't written your performance report for 2000, how do you write your performance plan for 2003? It's an issue of how often we have these reports. The law has an artificial timeline that doesn't fit any of us. The performance plan could extend over a longer period than a year. It can't possibly hit what you're doing in the next budget cycle. And we can't factor what we've learned in that report into the next cycle."

A representative of the General Accounting Office (GAO) suggested that it would be hard to change the requirement for annual reporting but that the agencies can specify what they are reporting annually by using an alternative reporting form. She said that GPRA is more flexible than agencies recognize.

Evaluating basic-research programs annually

Like the focus groups, this workshop featured extended discussions on the difficulty of evaluating the results of basic research each year. Such a requirement, several participants said, puts unrealistic pressure on a principal investigator to come up with the "next great discovery of the last 12 months." One participant noted that the "output" of good research is original knowledge, as measured by publications and perhaps new products, but that the "outcome" of that knowledge might be unknown for years.

As a result, DOD now looks at every research program not annually, but every 2 years. Review panels are asked whether adequate progress is being made toward stated goals.

NSF is planning to evaluate every basic-research program every 3 years, covering one-third of its portfolio every year. Thus, it is reporting on its programs every year, but reviewing a given program every 3 years. If there is an exciting discovery from a grant made 10-years previously, the discovery will be included after the year when it occurs.

A USDA representative pointed out that an evaluation is more effective when it comprehends several years: "you can tell your story better." But she said that evaluations can motivate people at the bench if they know that their results are expected and might be used by someone.

Verification and validation

Representatives of oversight bodies said that they would like more information about how agencies verify and validate their procedures for evaluation of research. For example, when an agency uses expert review to evaluate a program, who are the reviewers? How are they recruited? Are they all outside the agency? If not, when is the use of an internal panel justified? Are the users of research included in review panels? What qualifications are required? How are conflicts of interest avoided? What process do reviewers use? How good is the quality of the data that they are given? How much quantitative information is included?

The absence of such information from most GPRA reports leads to some suspicion about the objectivity of expert review and the independence of reviewers. Panelists agreed that although most oversight bodies do not want to review the curricula vitae of all an agency's reviewers, for example, they feel reassured if that material is at least described in reports and made available as necessary. It might be reasonable for a committee to undertake a sampling of a given agency's procedures, for example, to obtain a better understanding of the evaluation process. This topic has not been thoroughly discussed between agencies and oversight bodies, however.

Some participants indicated that the agencies themselves are best qualified to organize and validate their reporting procedures and that most of them have systems in place for doing so. The issue is not whether they should turn over the procedures or their validation to an outside body. They should simply be willing to describe the procedures in more detail than they do to provide a "sense of comfort" to the oversight bodies, which are primarily looking for an understanding of why agencies use the methods they use. "It's not a matter of right or wrong," said one oversight official. "It's just for us to understand what they did and why they did it that way."

Communication

Some agencies complained that oversight bodies issued conflicting requests, lacked consistency among personnel, failed to issue explicit guidelines, disliked new systems that were designed to comply with GPRA, and made unreasonable requests with regard to research activities, especially in large mission agencies.

Some oversight personnel complained that agencies did not explain the special needs of science adequately, did not reveal their specific planning and reporting mechanisms with sufficient transparency, did not adequately align the program activities with budget line items, and did not explain their validation and verification procedures for evaluating research programs.

One agency representative reported that he had never been contacted by a representative of Congress about GPRA. Some agency personnel were confused about why the appropriations committees did not seem to use or take an interest in agency GPRA reports. A congressional representative suggested that the level of aggregation of program activities was so high that committees could not understand or see the actual program activities.

A GAO representative stated, "We know what agencies are doing, but is it good, bad, or indifferent? What has worked, and what are the problems? Has it been a successful experiment?"

An EPA representative said that his agency had combined the budget with the performance plan, as suggested by GPRA, but that the "appropriators were upset that they weren't seeing what they were used to seeing" and asked for the old system. Other agency representatives reported the same difficulty with committees and sometimes with their own agencies.

One agency representative acknowledged that the process was still relatively young, and participants were still learning what the others wanted. "Three years ago," he said, "everyone was in denial that we were going to have to do anything with this."

Workshop Participant List
December 18-19, 2000

Panel Members:

Enriqueta C. Bond
President
The Burroughs Wellcome Fund
Research Triangle Park, North Carolina

Alan Schriesheim
Director Emeritus
Argonne National Laboratory
Argonne, Illinois

John E. Halver
Professor Emeritus in Nutrition
School of Aquatic and Fishery Sciences
University of Washington
Seattle, Washington

Wesley T. Huntress, Jr.
Director, Geophysical Laboratory
Carnegie Institution of Washington
Washington, D.C.

Louis J. Lanzerotti
Distinguished Member of the Technical Staff
Bell Laboratories, Lucent Technologies
Murray Hill, New Jersey

Rudolph A. Marcus
Arthur Amos Noyes Professor of Chemistry
California Institute of Technology
Pasadena, California

Stuart A. Rice [by confcall]
Frank P. Hixon Distinguished Service Professor
James Franck Institute
Department of Chemistry
The University of Chicago
Chicago, Illinois

Herbert H. Richardson
Associate Vice Chancellor of Engineering and Director, Texas Transportation Institute
The Texas A&M University System
College Station, Texas

Max D. Summers
Professor of Entomology
Department of Entomology
Texas A&M University
College Station, Texas

Morris Tanenbaum
Retired Vice Chairman and Chief Financial Officer, AT&T
Short Hills, New Jersey

Bailus Walker, Jr.
Professor of Environmental and Occupational Medicine
Howard University
Washington, D.C.

Robert M. White
University Professor and Director
Data Storage Systems Center
Carnegie Mellon University
Pittsburgh, Pennsylvania

Participants:

Mark Boroush
Office of Science Policy
Office of the Director
National Institutes of Health
Bethesda, Maryland

Howard Cantor
Performance Measurement Specialist
US Environmental Protection Agency
Washington, D.C.

Ann Carlson
Assistant to the NASA Chief Scientist
NASA Headquarters
Washington, D.C.

Beth Foster
Program Analyst
Office of the Deputy Under Secretary of Defense (S&T)
Washington, D.C.

Harriet Ganson
Chief, Planning, Evaluation and Legislation Branch of National Institute of Dental and Craniofacial Research, NIH
Bethesda, Maryland

Robin I. Kawazoe
Director, Office of Science Policy & Planning
National Institutes of Health
Bethesda, Maryland

Genevieve Knezo
Specialist, Science and Technology Policy
Congressional Research Service
Library of Congress
Washington, D.C.

Sara Mazie
US Department of Agriculture
Washington, D.C.

Jennifer McAndrew
Policy Analyst
National Institute of Standards and Technology
Gaithersburg, Maryland

Robin Nazzaro
Assistant Director
US General Accounting Office
Washington, D.C.

Nathaniel Pitts
Director, Office of Integrated Activities
The National Science Foundation
Arlington, Virginia

Joanne Spriggs
Deputy Under Secretary of Defense (S&T)
US Department of Defense
Washington, D.C.

John Uzzell
Director, Division of Evaluation, Office of Science Policy
National Institutes of Health
Rockville, Maryland

William J. Valdez
Director of the Office of Planning and Analysis
US Department of Energy
Washington, D.C.

EXECUTIVE SUMMARY OF *EVALUATING FEDERAL RESEARCH PROGRAMS: Research and the Government Performance and Results Act*

The Government Performance and Results Act (GPRA), enacted in 1993, focuses agency and oversight attention on the performance and results of government activities by requiring that all federal agencies measure and report on the results of their activities annually. Agencies are required to develop a strategic plan that sets goals and objectives for at least a 5-year period, an annual performance plan that translates the goals of the strategic plan into annual targets, and an annual performance report that demonstrates whether the targets are met. The Committee on Science, Engineering, and Public Policy (COSEPUP) of the National Academy of Sciences, the National Academy of Engineering, and the Institute of Medicine has addressed the issue of measuring and evaluating research in compliance with the requirements of GPRA.

COSEPUP recognizes the opportunities and challenges that GPRA presents for agencies that invest in research. GPRA offers those agencies the opportunity to communicate to policy-makers and the public the rationale for and results of their research programs. At the same time, GPRA presents substantial challenges to the agencies.

During the course of this study, COSEPUP held several workshops. In these workshops and in other input to the committee, we have heard two distinct and conflicting viewpoints on approaches to measuring basic research. One is that it should be possible to measure research, including basic research, annually and provide quantitative measures of the useful outcomes of both basic and applied research. The other is that, given the long-range nature of basic research, there is no sensible way to respond to the GPRA annual measurement requirement and that the best that can be done is to provide measures that appear to respond but in fact are essentially meaningless, such as a list of an agency's top 100 discoveries of the preceding year.

COSEPUP's view, spelled out in more detail in what follows, is different from both those viewpoints. In essence, our report takes two strong positions. First, the useful outcomes of basic research cannot be measured directly on an annual basis, because the usefulness of new basic knowledge is inherently too unpredictable; so the usefulness of basic research must be measured by historical reviews based on a much longer timeframe. Second, that does not mean that there are no meaningful measures of performance of basic research while the research is in progress; in fact, the committee believes that there are meaningful measures of quality, relevance, and leadership that are good predictors of eventual usefulness, that these measures can be reported regularly, and that they represent a sound way to ensure that the country is getting a good return on its basic research investment.

The problem of reporting on applied research is much simpler: it consists of systematically applying methods widely used in industry and in some parts of government. For example, an applied research program usually includes a series of milestones that should be achieved by particular times and a description of the intended final outcomes and their significance. Periodic reporting can indicate progress toward those milestones.

The remainder of this executive summary provides a more in-depth description of COSEPUP's conclusions and recommendations regarding how to evaluate federal research programs relative to GPRA. It also addresses coordination among federal research programs and human-resource issues. COSEPUP concludes that both basic research and applied research programs[1] can be meaningfully evaluated on a regular basis. For the applied research programs of the mission agencies, specific practical outcomes can be documented and progress toward their achievement can be measured annually. For example, if the Department of Energy adopted the goal of producing cheaper solar energy, it could measure the results of research directed toward decreasing the cost of solar cells; this applied research project would be evaluated annually against specific measurable milestones. However, the practical outcomes of basic research in science and engineering can seldom be identified while the research is in progress. Basic research has annual results that can be meaningfully evaluated, but these evaluations often do not give even a hint of ultimate practical outcomes.

History tells us unmistakably that by any measure, the benefit to the United States for leadership in basic research is extremely high—lives saved, inventions fostered, and jobs and wealth created. History also shows us how often basic research in science and engineering leads to outcomes that were unexpected or took many years or even decades to emerge. COSEPUP strongly believes that measures of the practical outcomes of basic research usually must be retrospective and historical and that the unpredictable nature of practical outcomes is an inherent and unalterable feature of basic research. For example, pre-World War II basic research on atomic structure contributed to today's Global Positioning System, an outcome of great practical and economic value, but, attempts to evaluate a year's worth of that early research even if they demonstrated high quality and world leadership, would have contained no hint of this particular outcome.

Since we cannot predict the ultimate practical outcomes of basic research, we must find ways to ensure that the basic research programs that the nation funds generate the kinds of knowledge that have given us great practical benefits in the past. To do that, we must find ways to measure the quality of our current research programs, their contributions to our world leadership in the relevant fields, and their relevance to agency goals and intended users.

World leadership is an important measure. In an earlier report (COSEPUP, 1993), COSEPUP recommended that, for the sake of the nation's well-being, the United States be among the leaders in all major fields of science and pre-eminent in selected fields of national importance. That is because a nation must be performing research at the forefront of a field if it is to understand, appropriate, and capitalize on current advances in that field, no matter where in the world they occur. New knowledge has value to nations where highly educated people performing cutting-edge research in the field of discovery can make use of the new knowledge when practical outcomes appear possible.

The people best qualified to evaluate basic or applied research are those with the knowledge and experience to understand its quality, and, in the case of applied research, its connection to public and agency goals. Evaluating basic research requires substantial scientific or engineering knowledge. Evaluating applied research requires, in addition, the ability to recognize its potential applicability to practical problems.

With those considerations in mind, COSEPUP has reached six conclusions and offers six recommendations regarding the evaluation of federally supported research programs.

Conclusion 1: Both applied research and basic research programs supported by the federal government can be evaluated meaningfully on a regular basis.

Conclusion 2: Agencies must evaluate their research programs by using measurements that match the character of the research. Differences in the character of the research will lead to differences in the appropriate timescale for measurement, in what is measurable and what is not, and in the expertise needed by those who contribute to the measurement process.

For applied research programs, progress toward specified practical outcomes can usually be measured annually by using milestones and other fairly standard approaches common in industry and in some parts of the federal government. For basic research, in contrast, progress toward practical outcomes cannot be measured annually, and attempts to measure such progress annually can in fact be harmful. Basic research progress can be reported annually in terms of quality, leadership, and relevance to agency goals, but practical outcomes can be measured only against a far longer historical perspective. In practical terms, because quality, leadership, and relevance will usually change slowly, the GPRA annual-reporting requirement can usually be met by minor updating of full evaluations that are done in a more flexible timeframe. There is a much greater chance that important events will take place in subfields, because of either scientific events or funding changes, so subprogram changes should constitute much of the updating.

Different expertise is required for measuring the worth of applied research and the worth of basic research. Measuring both requires technical and scientific knowledge, but applied research entails some factors that basic research does not, such as ultimate usability, so the input of potential users is required. That leads to our next conclusion.

Conclusion 3: The most effective means of evaluating federally funded research programs is expert review. Expert review—which includes quality review, relevance review, and benchmarking—should be used to assess both basic research and applied research programs.

Expert review is widely applied—used, for example, by congressional committees, by other professions, by industry boards, and throughout the realm of science and engineering—to answer complex questions through consultation with expert advisers. It is useful in helping an agency answer three kinds of questions of particular relevance to GPRA:

- What is the quality of the research program—for example, how good is current research work compared with other work being conducted in the field?[2] This question is best answered by reviewers who are sufficiently expert in the field being assessed to perform a *quality review*. This approach is traditionally called peer review. Peer review is commonly applied to projects, but here we are applying it to programs. The talent, objective judgment, and experience of these experts, or peers, are paramount and should be the criteria for their selection.
- Is the research program focused on the subjects most relevant to the agency mission? Another form of expert review is *relevance review*, in which potential users, joined by experts in related fields, evaluate the relevance of research to agency goals—is the research on subjects in which new understanding could be important in fulfilling the agency's mission? In reviewing the relevance of a program, a panel would assess the appropriateness of the direction of the research to the agency mission and its potential value to intended users.
- Is the research being performed at the forefront of scientific and technological knowledge? This is a relevant question for many programs, but it is particularly important for whole fields and subfields being supported. Evaluations of fields and subfields is best done through *international benchmarking* by a panel of experts who have sufficient stature and perspective to assess the international standing of research.

For agencies whose missions include a specific responsibility for basic research—such as the National Science Foundation in broad fields of science and engineering, the National Institutes of Health in fields related to health, or the Department of Energy in high-energy physics—world leadership in a field can itself be an agency goal. That is equally true for mission agencies, such as Department of Defense (DOD) but in more focused ways. For example, DOD can take as a goal world leadership in basic materials research relevant to its mission. Once such a goal is established, the usual measures of quality and leadership should be applied.

Conclusion 4: The nation cannot benefit from advances in science and technology without a continuing supply of well-educated and well-trained scientists and engineers. Without such a flow, the capability of an agency to fulfill its mission will be compromised. Agencies must pay increased attention to their human-resource requirements in terms of training and educating young scientists and engineers and in terms of providing an adequate supply of scientists and engineers to academe, industry, and federal laboratories.

Federal agencies that support research and exploit its results are able to do so because the education and training programs of the universities, in the course of performing much of that research, and the federal laboratories provide a continuing flow of qualified scientists and engineers. Even though section 1115(a)(3) of GPRA requires agencies to describe the human resources required to meet their performance goals, few agencies describe the importance of human resources or propose ways to ensure their adequacy in their strategic or performance plans.

Conclusion 5: Mechanisms for coordinating research programs in multiple agencies whose fields or subject matters overlap are insufficient.

It is common and valuable for agencies to approach similar fields of research from different perspectives. Indeed, this pluralism is a major strength of the U.S. research enterprise. But, better communication among agencies would enhance opportunities for collaboration, help keep important questions from being overlooked, and reduce instances of inefficient duplication of effort. Present mechanisms need strengthening.

Conclusion 6: The development of effective methods for evaluating and reporting performance requires the participation of the scientific and engineering community, whose members will necessarily be involved in expert review.

The researchers who work in agency, university, and industrial laboratories are the people who perform and best understand the research programs funded by the federal government. Many researchers contribute substantial time and effort to reviewing papers submitted for publication, grant applications, and program proposals, yet few of them are aware of GPRA, its objectives, and its mandates. Increased contact with and advice from the broader scientific and engineering community regarding the methods of determining and reporting quality and regarding the leadership position of agency research programs and the relevance of research to agency missions can benefit the GPRA process.

On the basis of those conclusions, COSEPUP offers the following recommendations:

Recommendation 1: Because both applied research and basic research can be evaluated meaningfully on a regular basis and are vital to research and mission agencies, research programs should be described in strategic and performance plans and evaluated in performance reports.

The performance of research is critical to the missions of many federal agencies. Therefore, a full description of an agency's goals and results, which is a principal objective of GPRA, must contain an evaluation of research activities and their relevance to the agency's mission.

Recommendation 2: For applied research programs, agencies should measure progress toward practical outcomes. For basic research programs, agencies should measure quality, relevance, and leadership. In addition, agencies should conduct periodic reviews of the overall practical outcomes of an agency's overall past support of applied and basic research. The use of measurements needs to recognize what can and cannot be measured. Misuse of measurement can lead to strongly negative results; for example, measuring basic research on the basis of short-term relevance would be extremely destructive to quality work.

Because the evaluation of applied research is directly connected to practical outcomes, whereas the evaluation of basic research is in terms of quality, relevance, and leadership, which ultimately lead to practical outcomes, there might be a tendency to bias an agency's overall research program toward applied research at the expense of basic research. This should be avoided, and a proper balance should be maintained.

Recommendation 3: Federal agencies should use expert review to assess the quality of research they support, the relevance of that research to their mission, and the leadership of the research. Expert review must strive for balance between having the most knowledgeable and the most independent individuals serve as members. Each agency should develop clear, explicit guidance with regard to structuring and employing expert review processes.

The most effective way to evaluate research programs is by expert review. The most commonly used form of expert review of quality is peer review. This operates on the premise that the people best qualified to judge the quality of research are experts in the field of research. This premise prevails across the research spectrum, from basic research to applied research. A second form of expert review is relevance review, in which potential users and experts in other fields or disciplines related to an agency's mission or to the potential application of the research evaluate the relevance of research to the agency's mission. A third form of expert review is benchmarking, in which an international panel of experts compares the level of leadership of a research program relative to research being performed worldwide.

Recommendation 4: Both research and mission agencies should describe in their strategic and performance plans the goal of developing and maintaining adequate human resources in fields critical to their missions both at the national level and in their agencies. Human resources should become a part of the evaluation of a research program along with the program's quality in terms of research advancement, relevance in terms of application development, and leadership in terms of the ability to take advantage of opportunities when they arise.

In early drafts of strategic and performance plans, agencies have generally omitted discussions of education and training, which are fundamental to the ability of agencies to fulfill their missions. The goal of developing and maintaining adequate human resources in fields critical to their missions should be supported by plans that produce that outcome. The nation cannot benefit from advances in science and technology without a continuing supply of well-educated and well-trained scientists and engineers. In addition, in the absence of such a flow, the capability of an agency to fulfill its mission will be compromised and the knowledge learned and technology developed will be lost.

Recommendation 5: Although GPRA is conducted agency-by-agency, a formal process should be established to identify and coordinate areas of research that are supported by multiple agencies. A lead agency should be identified for each field of research and that agency should be responsible for assuring that coordination occurs among the agencies.

It is common and valuable for multiple agencies to approach similar fields of research from different perspectives. Indeed, this pluralism is a major strength of the U.S. research enterprise. However, better communication among agencies would enhance opportunities for collaboration, help to keep important questions from being overlooked, and reduce instances of inefficient duplication of effort. A single agency should be identified to serve as the focal point for each particular field of research so that all significant supported fields are covered. Information regarding support for that field should be provided to all the agencies involved in it so that they can adjust their efforts to ensure that the field is appropriately covered. Agencies should use benchmarking, which affords the opportunity to look across fields, in their efforts to understand the status of a particular field of research.

Recommendation 6: The science and engineering community can and should play an important role in GPRA implementation. As a first step, they should become familiar with agency strategic and performance plans, which are available on the agencies' web sites.

The researchers who work in agency, university, and industrial laboratories are the people who perform and best understand the research programs funded by the federal government. Many researchers contribute substantial time and effort to reviewing papers submitted for publication, grant applications, and program proposals, but few of them are aware of GPRA. Their greater involvement in implementing GPRA would be beneficial to the country. Increased contact with and advice from the broader scientific and engineering community regarding both the quality and the leadership position of agency research programs and the relevance of the research to agency missions can benefit the GPRA process.

COSEPUP intends to address mechanisms and guidelines for implementing these recommendations in workshops and meetings with representatives from federal agencies, Congress, OMB, and oversight bodies. Given the diverse portfolio of research conducted by federal agencies and the urgency of addressing the question of how basic research can be evaluated in the context of GPRA, the level of detail and specificity needed in designing procedures and guidelines for implementation was beyond the scope of this report.

The Government Performance and Results Act provides an opportunity for the research community to ensure the effective use of the nation's research resources in meeting national needs and to articulate to policy-makers and the public the rationale for and results of research. We believe that our recommendations can assist federal agencies in complying with GPRA.

NOTES

1. For purposes of this study, *program* refers to a set of activities focused on a particular area that can include multiple projects with different risks, time horizons, and outcomes.

2. There are at least two aspects of quality—one absolute and one relative. The absolute aspects are related to the quality of the research plan, the methods by which it is being pursued, its role in education when conducted at a university, and the importance of its results to its sponsor, either obtained or expected. The relative aspects pertain to its leadership at the edge of an advancing field. Although the leadership aspect is generally important, the results might in some cases be of great importance to an agency albeit not at the leading edge of a field.

GOVERNMENT PERFORMANCE AND RESULTS ACT

S.20

One Hundred Third Congress
of the
United States of America
AT THE FIRST SESSION

Begun and held at the City of Washington on Tuesday, the fifth day of January, one thousand nine hundred and ninety-three

An Act

To provide for the establishment of strategic planning and performance measurement in the Federal Government, and for other purposes.

Be it enacted by the Senate and House of Representatives of the United States of America in Congress assembled,

SECTION 1. SHORT TITLE.

This Act may be cited as the "Government Performance and Results Act of 1993."

SEC. 2. FINDINGS AND PURPOSES.

(a) FINDINGS The Congress finds that—

(1) waste and inefficiency in Federal programs undermine the confidence of the American people in the Government

and reduces the Federal Government's ability to address adequately vital public needs;

(2) Federal managers are seriously disadvantaged in their efforts to improve program efficiency and effectiveness, because of insufficient articulation of program goals and inadequate information on program performance; and

(3) congressional policymaking, spending decisions and program oversight are seriously handicapped by insufficient attention to program performance and results.

(b) PURPOSES The purposes of this Act are to—

(1) improve the confidence of the American people in the capability of the Federal Government, by systematically holding Federal agencies accountable for achieving program results;

(2) initiate program performance reform with a series of pilot projects in setting program goals, measuring program performance against those goals, and reporting publicly on their progress;

(3) improve Federal program effectiveness and public accountability by promoting a new focus on results, service quality, and customer satisfaction;

(4) help Federal managers improve service delivery, by requiring that they plan for meeting program objectives and by providing them with information about program results and service quality;

(5) improve congressional decisionmaking by providing more objective information on achieving statutory objectives, and on the relative effectiveness and efficiency of Federal programs and spending; and

(6) improve internal management of the Federal Government.

SEC. 3. STRATEGIC PLANNING.

Chapter 3 of title 5, United States Code, is amended by adding after section 305 the following new section:

Sec. 306. Strategic plans

(a) No later than September 30, 1997, the head of each agency shall submit to the Director of the Office of Management and Budget and to the Congress a strategic plan for program activities. Such plan shall contain—

(1) a comprehensive mission statement covering the major functions and operations of the agency;

(2) general goals and objectives, including outcome-related goals and objectives, for the major functions and operations of the agency;

(3) a description of how the goals and objectives are to be achieved, including a description of the operational processes, skills and technology, and the human, capital, information, and other resources required to meet those goals and objectives;

(4) a description of how the performance goals included in the plan required by section 1115(a) of title 31 shall be related to the general goals and objectives in the strategic plan;

(5) an identification of those key factors external to the agency and beyond its control that could significantly affect the achievement of the general goals and objectives; and

(6) a description of the program evaluations used in establishing or revising general goals and objectives, with a schedule for future program evaluations.

(b) The strategic plan shall cover a period of not less than five years forward from the fiscal year in which it is submitted, and shall be updated and revised at least every three years.

(c) The performance plan required by section 1115 of title 31 shall be consistent with the agency's strategic plan. A performance plan may not be submitted for a fiscal year not covered by a current strategic plan under this section.

(d) When developing a strategic plan, the agency shall consult with the Congress, and shall solicit and consider the views and suggestions of those entities potentially affected by or interested in such a plan.

(e) The functions and activities of this section shall be considered to be inherently Governmental functions. The drafting of strategic plans under this section shall be performed only by Federal employees.

(f) For purposes of this section the term 'agency' means an Executive agency defined under section 105, but does not include the Central Intelligence Agency, the General Accounting Office, the Panama Canal Commission, the United States Postal Service, and the Postal Rate Commission.

SEC. 4. ANNUAL PERFORMANCE PLANS AND REPORTS.

(a) BUDGET CONTENTS AND SUBMISSION TO CONGRESS Section 1105(a) of title 31, United States Code, is amended by adding at the end thereof the following new paragraph:

> (29) beginning with fiscal year 1999, a Federal Government performance plan for the overall budget as provided for under section 1115.

(b) PERFORMANCE PLANS AND REPORTS Chapter 11 of title 31, United States Code, is amended by adding after section 1114 the following new sections:

Sec. 1115. Performance plans

(a) In carrying out the provisions of section 1105(a)(29), the Director of the Office of Management and Budget shall require each agency to prepare an annual performance plan covering each program activity set forth in the budget of such agency. Such plan shall—

(1) establish performance goals to define the level of performance to be achieved by a program activity;

(2) express such goals in an objective, quantifiable, and measurable form unless authorized to be in an alternative form under subsection (b);

(3) briefly describe the operational processes, skills and technology, and the human, capital, information, or other resources required to meet the performance goals;

(4) establish performance indicators to be used in measuring or assessing the relevant outputs, service levels, and outcomes of each program activity;

(5) provide a basis for comparing actual program results with the established performance goals; and

(6) describe the means to be used to verify and validate measured values.

(b) If an agency, in consultation with the Director of the Office of Management and Budget, determines that it is not feasible to express the performance goals for a particular program activity in an objective, quantifiable, and measurable form, the Director of the Office of Management and Budget may authorize an alternative form. Such alternative form shall—

(1) include separate descriptive statements of—

(A)(i) a minimally effective program, and

(ii) a successful program, or

(B) such alternative as authorized by the Director of the Office of Management and Budget, with sufficient precision and in such terms that would allow for an accurate, independent determination of whether the program activity's performance meets the criteria of the description; or

(2) state why it is infeasible or impractical to express a performance goal in any form for the program activity.

(c) For the purpose of complying with this section, an agency may aggregate, disaggregate, or consolidate program activities, except that any aggregation or consolidation may not omit or minimize the significance of any program activity constituting a major function or operation for the agency.

(d) An agency may submit with its annual performance plan an appendix covering any portion of the plan that—

(1) is specifically authorized under criteria established by an Executive order to be kept secret in the interest of national defense or foreign policy; and

(2) is properly classified pursuant to such Executive order.

(e) The functions and activities of this section shall be considered to be inherently Governmental functions. The drafting of performance plans under this section shall be performed only by Federal employees.

(f) For purposes of this section and sections 1116 through 1119, and sections 9703 and 9704 the term—

(1) "agency" has the same meaning as such term is defined under section 306(f) of title 5;

(2) "outcome measure" means an assessment of the results of a program activity compared to its intended purpose;

(3) "output measure" means the tabulation, calculation, or

recording of activity or effort and can be expressed in a quantitative or qualitative manner;

(4) "performance goal" means a target level of performance expressed as a tangible, measurable objective, against which actual achievement can be compared, including a goal expressed as a quantitative standard, value, or rate;

(5) "performance indicator" means a particular value or characteristic used to measure output or outcome;

(6) "program activity" means a specific activity or project as listed in the program and financing schedules of the annual budget of the United States Government; and

(7) "program evaluation" means an assessment, through objective measurement and systematic analysis, of the manner and extent to which Federal programs achieve intended objectives.

Sec. 1116. Program performance reports

(a) No later than March 31, 2000, and no later than March 31 of each year thereafter, the head of each agency shall prepare and submit to the President and the Congress, a report on program performance for the previous fiscal year.

(b)(1) Each program performance report shall set forth the performance indicators established in the agency performance plan under section 1115, along with the actual program performance achieved compared with the performance goals expressed in the plan for that fiscal year.

(2) If performance goals are specified in an alternative form under section 1115(b), the results of such program shall be described in relation to such specifications, including whether the performance failed to meet the criteria of a minimally effective or successful program.

(c) The report for fiscal year 2000 shall include actual results for the preceding fiscal year, the report for fiscal year 2001 shall include actual results for the two preceding fiscal years, and the report for fiscal year 2002 and all subsequent reports shall include actual results for the three preceding fiscal years.

(d) Each report shall—

(1) review the success of achieving the performance goals of the fiscal year;

(2) evaluate the performance plan for the current fiscal year relative to the performance achieved toward the performance goals in the fiscal year covered by the report;

(3) explain and describe, where a performance goal has not been met (including when a program activity's performance is determined not to have met the criteria of a successful program activity under section 1115(b)(1)(A)(ii) or a corresponding level of achievement if another alternative form is used)—

(A) why the goal was not met;

(B) those plans and schedules for achieving the established performance goal; and

(C) if the performance goal is impractical or infeasible, why that is the case and what action is recommended;

(4) describe the use and assess the effectiveness in achieving performance goals of any waiver under section 9703 of this title; and

(5) include the summary findings of those program evaluations completed during the fiscal year covered by the report.

(e) An agency head may include all program performance information required annually under this section in an annual financial statement required under section 3515 if any such statement is submitted to the Congress no later than March 31 of the applicable fiscal year.

(f) The functions and activities of this section shall be considered to be inherently Governmental functions. The drafting of program performance reports under this section shall be performed only by Federal employees.

Sec. 1117. Exemption

The Director of the Office of Management and Budget may exempt from the requirements of sections 1115 and 1116 of this title and section 306 of title 5, any agency with annual outlays of $20,000,000 or less.

SEC. 5. MANAGERIAL ACCOUNTABILITY AND FLEXIBILITY.

(a) MANAGERIAL ACCOUNTABILITY AND FLEXIBILITY Chapter 97 of title 31, United States Code, is amended by adding after section 9702, the following new section:

Sec. 9703. Managerial accountability and flexibility

(a) Beginning with fiscal year 1999, the performance plans required under section 1115 may include proposals to waive administrative procedural requirements and controls, including specification of personnel staffing levels, limitations on compensation or remuneration, and prohibitions or restrictions on funding transfers among budget object classification 20 and subclassifications 11, 12, 31, and 32 of each annual budget submitted under section 1105, in return for specific individual or organization accountability to achieve a performance goal. In preparing and submitting the performance plan under section 1105(a)(29), the Director of the Office of Management and

Budget shall review and may approve any proposed waivers. A waiver shall take effect at the beginning of the fiscal year for which the waiver is approved.

(b) Any such proposal under subsection (a) shall describe the anticipated effects on performance resulting from greater managerial or organizational flexibility, discretion, and authority, and shall quantify the expected improvements in performance resulting from any waiver. The expected improvements shall be compared to current actual performance, and to the projected level of performance that would be achieved independent of any waiver.

(c) Any proposal waiving limitations on compensation or remuneration shall precisely express the monetary change in compensation or remuneration amounts, such as bonuses or awards, that shall result from meeting, exceeding, or failing to meet performance goals.

(d) Any proposed waiver of procedural requirements or controls imposed by an agency (other than the proposing agency or the Office of Management and Budget) may not be included in a performance plan unless it is endorsed by the agency that established the requirement, and the endorsement included in the proposing agency's performance plan.

(e) A waiver shall be in effect for one or two years as specified by the Director of the Office of Management and Budget in approving the waiver. A waiver may be renewed for a subsequent year. After a waiver has been in effect for three consecutive years, the performance plan prepared under section 1115 may propose that a waiver, other than a waiver of limitations on compensation or remuneration, be made permanent.

(f) For purposes of this section, the definitions under section 1115(f) shall apply.

SEC. 6. PILOT PROJECTS.

(a) PERFORMANCE PLANS AND REPORTS Chapter 11 of title 31, United States Code, is amended by inserting after section 1117 (as added by section 4 of this Act) the following new section:

Sec. 1118. Pilot projects for performance goals

(a) The Director of the Office of Management and Budget, after consultation with the head of each agency, shall designate not less than ten agencies as pilot projects in performance measurement for fiscal years 1994, 1995, and 1996. The selected agencies shall reflect a representative range of Government functions and capabilities in measuring and reporting program performance.

(b) Pilot projects in the designated agencies shall undertake the preparation of performance plans under section 1115, and program performance reports under section 1116, other than section 1116(c), for one or more of the major functions and operations of the agency. A strategic plan shall be used when preparing agency performance plans during one or more years of the pilot period.

(c) No later than May 1, 1997, the Director of the Office of Management and Budget shall submit a report to the President and to the Congress which shall—

(1) assess the benefits, costs, and usefulness of the plans and reports prepared by the pilot agencies in meeting the purposes of the Government Performance and Results Act of 1993;

(2) identify any significant difficulties experienced by the pilot agencies in preparing plans and reports; and

(3) set forth any recommended changes in the require-

ments of the provisions of Government Performance and Results Act of 1993, section 306 of title 5, sections 1105, 1115, 1116, 1117, 1119 and 9703 of this title, and this section.

(b) MANAGERIAL ACCOUNTABILITY AND FLEXIBILITY Chapter 97 of title 31, United States Code, is amended by inserting after section 9703 (as added by section 5 of this Act) the following new section:

Sec. 9704. Pilot projects for managerial accountability and flexibility

(a) The Director of the Office of Management and Budget shall designate not less than five agencies as pilot projects in managerial accountability and flexibility for fiscal years 1995 and 1996. Such agencies shall be selected from those designated as pilot projects under section 1118 and shall reflect a representative range of Government functions and capabilities in measuring and reporting program performance.

(b) Pilot projects in the designated agencies shall include proposed waivers in accordance with section 9703 for one or more of the major functions and operations of the agency.

(c) The Director of the Office of Management and Budget shall include in the report to the President and to the Congress required under section 1118(c)—

(1) an assessment of the benefits, costs, and usefulness of increasing managerial and organizational flexibility, discretion, and authority in exchange for improved performance through a waiver; and

(2) an identification of any significant difficulties experienced by the pilot agencies in preparing proposed waivers.

(d) For purposes of this section the definitions under section 1115(f) shall apply.

(c) PERFORMANCE BUDGETING Chapter 11 of title 31, United States Code, is amended by inserting after section 1118 (as added by section 6 of this Act) the following new section:

Sec. 1119. Pilot projects for performance budgeting

(a) The Director of the Office of Management and Budget, after consultation with the head of each agency shall designate not less than five agencies as pilot projects in performance budgeting for fiscal years 1998 and 1999. At least three of the agencies shall be selected from those designated as pilot projects under section 1118, and shall also reflect a representative range of Government functions and capabilities in measuring and reporting program performance.

(b) Pilot projects in the designated agencies shall cover the preparation of performance budgets. Such budgets shall present, for one or more of the major functions and operations of the agency, the varying levels of performance, including outcome-related performance, that would result from different budgeted amounts.

(c) The Director of the Office of Management and Budget shall include, as an alternative budget presentation in the budget submitted under section 1105 for fiscal year 1999, the performance budgets of the designated agencies for this fiscal year.

(d) No later than March 31, 2001, the Director of the Office of Management and Budget shall transmit a report to the President and to the Congress on the performance budgeting pilot projects which shall—

(1) assess the feasibility and advisability of including a performance budget as part of the annual budget submitted under section 1105;

(2) describe any difficulties encountered by the pilot agencies in preparing a performance budget;

(3) recommend whether legislation requiring performance budgets should be proposed and the general provisions of any legislation; and

(4) set forth any recommended changes in the other requirements of the Government Performance and Results Act of 1993, section 306 of title 5, sections 1105, 1115, 1116, 1117, and 9703 of this title, and this section.

(e) After receipt of the report required under subsection (d), the Congress may specify that a performance budget be submitted as part of the annual budget submitted under section 1105.

SEC. 7. UNITED STATES POSTAL SERVICE.

Part III of title 39, United States Code, is amended by adding at the end thereof the following new chapter:

CHAPTER 28—STRATEGIC PLANNING AND PERFORMANCE MANAGEMENT

Sec.

2801. Definitions.
2802. Strategic plans.
2803. Performance plans.
2804. Program performance reports.
2805. Inherently Governmental functions.

Sec. 2801. Definitions

For purposes of this chapter the term—

(1) "outcome measure" refers to an assessment of the results of a program activity compared to its intended purpose;

(2) "output measure" refers to the tabulation, calculation, or recording of activity or effort and can be expressed in a quantitative or qualitative manner;

(3) "performance goal" means a target level of performance expressed as a tangible, measurable objective, against which actual achievement shall be compared, including a goal expressed as a quantitative standard, value, or rate;

(4) "performance indicator" refers to a particular value or characteristic used to measure output or outcome;

(5) "program activity" means a specific activity related to the mission of the Postal Service; and

(6) "program evaluation" means an assessment, through objective measurement and systematic analysis, of the manner and extent to which Postal Service programs achieve intended objectives.

Sec. 2802. Strategic plans

(a) No later than September 30, 1997, the Postal Service shall submit to the President and the Congress a strategic plan for its program activities. Such plan shall contain—

(1) a comprehensive mission statement covering the major functions and operations of the Postal Service;

(2) general goals and objectives, including outcome-related goals and objectives, for the major functions and operations of the Postal Service;

(3) a description of how the goals and objectives are to be achieved, including a description of the operational processes, skills and technology, and the human, capital, information, and other resources required to meet those goals and objectives;

(4) a description of how the performance goals included in

the plan required under section 2803 shall be related to the general goals and objectives in the strategic plan;

(5) an identification of those key factors external to the Postal Service and beyond its control that could significantly affect the achievement of the general goals and objectives; and

(6) a description of the program evaluations used in establishing or revising general goals and objectives, with a schedule for future program evaluations.

(b) The strategic plan shall cover a period of not less than five years forward from the fiscal year in which it is submitted, and shall be updated and revised at least every three years.

(c) The performance plan required under section 2803 shall be consistent with the Postal Service's strategic plan. A performance plan may not be submitted for a fiscal year not covered by a current strategic plan under this section.

(d) When developing a strategic plan, the Postal Service shall solicit and consider the views and suggestions of those entities potentially affected by or interested in such a plan, and shall advise the Congress of the contents of the plan.

Sec. 2803. Performance plans

(a) The Postal Service shall prepare an annual performance plan covering each program activity set forth in the Postal Service budget, which shall be included in the comprehensive statement presented under section 2401(g) of this title. Such plan shall—

(1) establish performance goals to define the level of performance to be achieved by a program activity;

(2) express such goals in an objective, quantifiable, and measurable form unless an alternative form is used under

subsection (b);

(3) briefly describe the operational processes, skills and technology, and the human, capital, information, or other resources required to meet the performance goals;

(4) establish performance indicators to be used in measuring or assessing the relevant outputs, service levels, and outcomes of each program activity;

(5) provide a basis for comparing actual program results with the established performance goals; and

(6) describe the means to be used to verify and validate measured values.

(b) If the Postal Service determines that it is not feasible to express the performance goals for a particular program activity in an objective, quantifiable, and measurable form, the Postal Service may use an alternative form. Such alternative form shall—

(1) include separate descriptive statements of—

(A) a minimally effective program, and

(B) a successful program,

with sufficient precision and in such terms that would allow for an accurate, independent determination of whether the program activity's performance meets the criteria of either description; or

(2) state why it is infeasible or impractical to express a performance goal in any form for the program activity.

(c) In preparing a comprehensive and informative plan under this section, the Postal Service may aggregate, disaggregate, or consolidate program activities, except that any aggregation or consolidation may not omit or minimize the significance of any program activity constituting a major function or operation.

(d) The Postal Service may prepare a non-public annex to its plan covering program activities or parts of program activities relating to—

(1) the avoidance of interference with criminal prosecution; or

(2) matters otherwise exempt from public disclosure under section 410(c) of this title.

Sec. 2804. Program performance reports

(a) The Postal Service shall prepare a report on program performance for each fiscal year, which shall be included in the annual comprehensive statement presented under section 2401(g) of this title.

(b)(1) The program performance report shall set forth the performance indicators established in the Postal Service performance plan, along with the actual program performance achieved compared with the performance goals expressed in the plan for that fiscal year.

(2) If performance goals are specified by descriptive statements of a minimally effective program activity and a successful program activity, the results of such program shall be described in relationship to those categories, including whether the performance failed to meet the criteria of either category.

(c) The report for fiscal year 2000 shall include actual results for the preceding fiscal year, the report for fiscal year 2001 shall include actual results for the two preceding fiscal years, and the report for fiscal year 2002 and all subsequent reports shall include actual results for the three preceding fiscal years.

(d) Each report shall—

(1) review the success of achieving the performance goals of the fiscal year;

(2) evaluate the performance plan for the current fiscal year relative to the performance achieved towards the performance goals in the fiscal year covered by the report;

(3) explain and describe, where a performance goal has not been met (including when a program activity's performance is determined not to have met the criteria of a successful program activity under section 2803(b)(2))—

(A) why the goal was not met;

(B) those plans and schedules for achieving the established performance goal; and

(C) if the performance goal is impractical or infeasible, why that is the case and what action is recommended; and

(4) include the summary findings of those program evaluations completed during the fiscal year covered by the report.

Sec. 2805. Inherently Governmental functions

The functions and activities of this chapter shall be considered to be inherently Governmental functions. The drafting of strategic plans, performance plans, and program performance reports under this section shall be performed only by employees of the Postal Service.

SEC. 8. CONGRESSIONAL OVERSIGHT AND LEGISLATION.

(a) IN GENERAL Nothing in this Act shall be construed as limiting the ability of Congress to establish, amend, suspend, or annul a performance goal. Any such action shall have the effect of superseding that goal in the plan submitted under section

1105(a)(29) of title 31, United States Code.

(b) GAO REPORT No later than June 1, 1997, the Comptroller General of the United States shall report to Congress on the implementation of this Act, including the prospects for compliance by Federal agencies beyond those participating as pilot projects under sections 1118 and 9704 of title 31, United States Code.

SEC. 9. TRAINING.

The Office of Personnel Management shall, in consultation with the Director of the Office of Management and Budget and the Comptroller General of the United States, develop a strategic planning and performance measurement training component for its management training program and otherwise provide managers with an orientation on the development and use of strategic planning and program performance measurement.

SEC. 10. APPLICATION OF ACT.

No provision or amendment made by this Act may be construed as—

(1) creating any right, privilege, benefit, or entitlement for any person who is not an officer or employee of the United States acting in such capacity, and no person who is not an officer or employee of the United States acting in such capacity shall have standing to file any civil action in a court of the United States to enforce any provision or amendment made by this Act; or

(2) superseding any statutory requirement, including any requirement under section 553 of title 5, United States Code.

SEC. 11. TECHNICAL AND CONFORMING AMENDMENTS.

(a) AMENDMENT TO TITLE 5, UNITED STATES CODE The table of sections for chapter 3 of title 5, United States Code, is amended by adding after the item relating to section 305 the following:

306. Strategic plans.

(b) AMENDMENTS TO TITLE 31, UNITED STATES CODE

(1) AMENDMENT TO CHAPTER 11 The table of sections for chapter 11 of title 31, United States Code, is amended by adding after the item relating to section 1114 the following:

1115. Performance plans.

1116. Program performance reports.

1117. Exemptions.

1118. Pilot projects for performance goals.

1119. Pilot projects for performance budgeting.

(2) AMENDMENT TO CHAPTER 97 The table of sections for chapter 97 of title 31, United States Code, is amended by adding after the item relating to section 9702 the following:

9703. Managerial accountability and flexibility.

9704. Pilot projects for managerial accountability and flexibility.

(c) AMENDMENT TO TITLE 39, UNITED STATES CODE The table of chapters for part III of title 39, United

States Code, is amended by adding at the end thereof the following new item:

2801.

Speaker of the House of Representatives.

Vice President of the United States and

President of the Senate.

FEDERAL AGENCY GPRA WEB SITES

Government Wide Performance Plan
http://clinton4.nara.gov/textonly/OMB/mgmt-gpra/gpptoc.html

General Accounting Office – Special Publications and Software
http://www.gao.gov/special.pubs/z3publist.htm (last updated June, '98)

OMB Watch (Office of Management and Budget)
http://www.gao.gov/special.pubs/z3publist.htm

Office of Management and Budget (OMB) and GPRA (1993)
http://www.gao.gov/special.pubs/z3publist.htm

GovExec.com GPRA and Results
http://www.govexec.com/gpra/

Department of Agriculture (USDA)
Strategic Plan: http://www.usda.gov/ocfo/strat/index1.htm
Performance Plan: http://www.usda.gov/ocfo/advcncl/accontnt.html

Department of Defense
Strategic Plan: http://www.defenselink.mil/pubs/qdr/
Performance Plan: http://www.dtic.mil/execsec/adr98/apdx_j.html

Department of Energy
Strategic Plan: http://www.cfo.doe.gov/stratmgt/plan/DOE-SP-full2.pdf
Performance Plan: http://www.cfo.doe.gov/stratmgt/DOE00rpt.pdf

Office of Health and Human Services
Strategic Plan: http://www.hhs.gov/ocr/ocr2001-22.html
Performance Plan: http://www.hhs.gov/ocr/reserv2.htm

Department of Transportation
Strategic Plan: http://stratplan.dot.gov/
Performance Plan: http://ostpxweb.dot.gov/budget/Perfplan02.pdf

Environmental Protection Agency
Strategic Plan: http://www.epa.gov/ocfopage/plan/plan.htm
Performance Plan: http://www.epa.gov/ocfo/budget/2kplan/2000contents.htm

National Aeronautics and Space Administration (NASA)
Strategic Plan: http://www.hq.nasa.gov/office/nsp/

National Oceanographic and Atmospheric Administration

Strategic Plan: http://www.noaa.gov/str-plan/toc.htm

National Institute of Standards and Technology

Strategic Plan: http://www.nist.gov/director/planning/strategicplanning.htm
Performance Plan: http://www.nist.gov/hearings/1997/sgpra624.htm

National Science Foundation

Strategic Plan: http://www.nsf.gov/od/gpraplan/gpraplan.htm
Performance Plan: http://www.nsf.gov/od/gpra/perfplan/fy2001/perfplanfinal.html

US Geological Survey

Strategic Plan: http://www.usgs.gov/stratplan/
Performance Plan: http://access.usgs.gov/gpra_goals99.doc